## Statistical Survey Design and Evaluating Impact

Sample surveys are a popular and viable alternative to the official statistics and censuses to collect descriptive information on population characteristics. Sampling designs that enable the selection of a specific portion of the population have been regarded as an effective way to provide reliable and valid estimates of the population parameters.

Two types of statistical designs – sample survey and evaluation – are pivotal tools for research in different disciplines. Sample survey designs focus on representation to capture important characteristics of a population. Evaluation designs emphasize on making the two groups, treatment and control, look as similar as possible to each other in terms of the covariates. Evaluation designs strive for internal validity whereas sample survey designs try to achieve external validity.

Statistical designs are critical in the evaluation of social programmes and policies and in exploring causal relationships. This book discusses some important methodologies to design sample surveys and impact evaluation. Solved examples are included to illustrate each technique discussed in the book. Case studies from around the world highlight applications of the techniques. Sources of biases and ways to overcome them have been outlined in a clear and concise manner. Clear guidelines with remedial measures have been outlined to facilitate choosing a suitable sampling design from among many alternatives.

**Tarun Kumar Roy** is former Director, International Institute for Population Sciences (IIPS), Mumbai, India. He taught courses on evaluation of family welfare programs and sampling design. He was engaged in the National Family Health Survey, India and also serves as an expert on the Sampling Review Committee for the Global Adult Tobacco Survey (GATS), USA since 2008.

**Rajib Acharya** is an Associate with the Population Council, New Delhi. He is a statistician and demographer with more than 16 years of experience in designing large sample surveys, sample size determination, processing and managing large scale cross-sectional and longitudinal data sets and analysis of survey data on population, health and nutrition.

**Arun Kumar Roy** is Chief Executive of Economic Information Technology, a socio-economic research organization in Kolkata. He has served as a Monitoring and Evaluation Consultant to several organizations including the World Bank in Odisha Institutional Council on Management of Population Programs, Malaysia and German Agro Action (GAA), Bonn.

# Statistical Survey Design and Evaluating Impact

Tarun Kumar Roy
Rajib Acharya
Arun Kumar Roy

CAMBRIDGE
UNIVERSITY PRESS

CAMBRIDGE
UNIVERSITY PRESS

4843/24, 2nd Floor, Ansari Road, Daryaganj, Delhi - 110002, India

Cambridge University Press is part of the University of Cambridge.

It furthers the University's mission by disseminating knowledge in the pursuit of education, learning and research at the highest international levels of excellence.

www.cambridge.org

Information on this title: www.cambridge.org/9781107146457



First published 2016

Printed in India by Shree Maitrey Printech Pvt. Ltd., Noida

*A catalogue record for this publication is available from the British Library*

ISBN 978-1-107-14645-7 Hardback

*To our parents for their invaluable support*

# Contents

# Figures

# Tables

# Foreword

When Dr Tarun Kumar Roy received a major award for excellence in teaching, on the occasion of the 50th (jubilee) anniversary of the International Institute for Population Sciences in Mumbai, where he was a previous director, the applause from the large audience was thunderous and sustained. It went on for more than ten minutes. The atmosphere was exhilarating and the good will was infectious. I had never before seen such an enthusiastic display of affection for a professor and a scholar, but having worked with Dr Roy for more than twenty years I was not surprised. What causes an individual to garner such respect from a very diverse audience? Dr Roy is the consummate professor and a true guru in his field. His ability to connect with students and researchers on both a cerebral and personal level is a key factor. He is well known for the depth of his knowledge in numerous areas (including statistical survey design, demography, international development, and impact evaluation). His ability to accurately answer questions from students and colleagues in his areas of expertise in simple terms, and his honesty in not being afraid to admit when he does not know an answer, which thankfully is a rare event, has always been highly appreciated. Personally, I have worked on large-scale nationally representative household surveys in 17 low- and middle-income countries, and in all that time I have rarely come across a colleague who is as knowledgeable, dependable, and humble as Professor Roy.

Together with his distinguished co-authors (Rajib Acharya and A. K. Roy), T. K. Roy has managed to translate his ability to connect with students and faculty on multiple levels into an immensely readable book that will be equally useful to beginners and seasoned professionals. The authors' skill in getting the material across to readers comes not only from their solid background in scientific survey sampling methods and statistics but also from their on-the-ground experience with some of the largest and most complex household surveys that have ever been conducted, including four National Family Health Surveys in India (that have included interviews with a total of more than one million respondents). This experience has allowed the authors to go well beyond the theory of sampling statistics to the application of those statistics to real-life situations.

To borrow a phrase from Amitav Banerjee, this book embodies the principle of "statistics without tears". It is both comprehensive and accessible. It covers the gamut from basic sample designs to designs for causal effects. The book also includes chapters on tackling non-sampling errors and statistical tests for measuring impact, that readers will find invaluable. Any discussion of sampling survey design would be incomplete without the attention that this book pays to problems that are likely to be encountered in designing and implementing surveys. The clear, down-to-earth examples that accompany key topics and the more in-depth case studies will enhance the reader's understanding of the concepts and statistical equations.

This volume will serve as a valuable reference that any researcher, survey planner or student needs to help understand survey statistics, plan high-quality surveys, and make informed use of survey data. It is also an essential compendium that can be relied on to provide answers to a variety of theoretical and practical questions relating to statistical survey design and impact evaluation. The authors' decision to cover such a wide variety of topics in a single volume and to speak to a variety of different audiences will certainly make this book one of the most useful sources of information on survey statistics that has been published in recent years.

Fred Arnold
The Demographic and Health Surveys Programme
Senior Fellow
ICF International
Rockville, Maryland, USA

# Preface

The twenty-first century is the *Century of Information*. Scientifically designed gathering, collation, analyses, interpretation, sharing and dissemination of information have become the primary occupation of millions, as the activity encompasses the widest possible spectrum of subjects ranging from education to development including social sciences, developmental economics and planning, medical sciences, engineering and even commercial subjects such as marketing and branding. The two major pillars in this grand edifice are 'survey designs' and 'evaluation designs'. In spite of the significant potential of statistical designs in the development of valid and reliable information, there is a wide and visible gap between development of theory and its practice.

Professor Leslie Kish, one of the world's top survey statisticians, once commented in his paper 'The Hundred Years' Wars of Survey Sampling', "my central complaint is that over 95% of statistical attention in academia, textbooks and publications is devoted to mathematical statistical analysis and only 2% to design" (Kish, 2003a). With regard to attention to designs, his comment is as apt as ever. This is a rather unfortunate situation, particularly since the two designs can compliment and nurture each other's application and growth. Professor Kish continued to remark "the consequences of that neglect are too often poor designs by non-statisticians (engineers, economists etc.)". Given this, the present book is a modest attempt to provide practitioners with tools for application of both designs, irrespective of their field of expertise.

## A BRIEF HISTORY OF TWO DESIGNS

Understanding the prevalence of variables and their causal relationships has long been the prime focus of research. Initially, the interest was on obtaining a count of the population, such as adult population for tax purposes as well as for enrolment in the military, availability of land for habitation and agriculture etc. Graunt (1662) estimated the population of London around the year 1662. Utilising parish registers, he first estimated the ratio of number of burials per family. In fact, he observed that there would be 3 burials per 11 families in a year. He also obtained the total number of burials in London. With the help of the total and the ratio, he then estimated the total number of families. An assumption about the average family size then resulted in the estimated total population of the country (Bethlehem, 2009). Laplace used a type of survey to estimate the population of France in around the year 1812. Although he used an approach similar to Graunt's, his method of estimation of the ratio (the ratio used by him was based on average number of persons per department, an aerial unit) was more systematic and based on selection of a number of areas in the country. These methods can be called *inductive reasoning*, not sampling.

Survey sampling began towards the end of the nineteenth century. During that time and till the 1930s, both probability and purposive sampling, particularly the latter, were in use and there was no basis for providing any idea about the likely error or precision of an estimate. Opinion poll surveys were used by Literary Digest, a news magazine in US between 1890 and 1938. The surveys successfully predicted the presidential candidates in a few elections, but failed to do so in the 1936 U.S. election. The sample size considered for this survey (1936 election) was quite large; more than two million persons participated in it. However, the selection was done purposively. Most respondents for the opinion poll were republicans; majority members of Literary Digest. This fact was not taken into account during the analysis, with the result that the estimate of a republican victory was grossly biased.

Work by Neyman (1934) provided the basis for modern probability sampling. He outlined a procedure for calculating the error of an estimate drawn from a probability sample and he also demonstrated the pitfalls in selection of a purposive sample (also see O'Muircheartaigh and Wong, 1981).

Studies on causal relationship, prior to the 930s, were largely concentrated on obtaining an average value of a dependent variable in the presence and absence of an independent variable. For example, in the 1920s an increase in lung cancer cases was noticed. Studies were undertaken to examine the effect of smoking on the disease. The output variable, proportion of persons suffering from the disease was compared among smokers and non-smokers. Early application of randomised experiments was largely in the field of agriculture, which began in the late 1920s after the pioneering works of Fisher (1918, 1925) and Neyman (1923).

## WHY THIS BOOK AND HOW IT IS ORGANIZED

There are two major purposes behind gathering information in the field of social sciences. The first is to estimate prevalence or level of a particular population characteristic and the other is to understand its dispersion in the population. Estimating prevalence can encompass understanding levels of various socio-economic characteristics of a population like education, school attendance among children, household income, occupation etc. or about health related indicators such as prevalence of a particular disease, disability, death or availability of health facilities.

Some of this data is routinely collected by official registration systems, while some other is availed through censuses conducted periodically. Of late, sample surveys have become a viable and important tool for obtaining a variety of information that is required for planning purposes or to serve one time-specific research purposes. In essence, a sample survey entails selection of a portion (sample) from a population so that the collected information from the sample would permit one to infer or generalize about the population. For example, a small sample consisting of a few thousands households would suffice to provide valid information on many of the above for a populace consisting of several millions households.

However estimates obtained from a sample survey are affected by two broad types of errors — the sampling and the non-sampling error. The sampling error occurs because it provides an estimate of a population parameter based on a sample of population units and is not based on all the units. The non-sampling error, on the other hand, occurs due to reasons other than sampling only a few units from the whole population such as extent of training of the surveyors, strictness with which standard survey procedures were followed, interviewer bias or even bias by respondents. Non-sampling errors are even present in the complete enumeration of the population. With significant cost advantage, an appropriately designed sample survey is often the most preferred option to collect data. In fact, it may be that the total error consisting of both sampling and non-sampling errors in a sample estimate is less than that in the complete enumeration. The demand for sample surveys is continuously increasing as countries get modernized and progress economically. The discussion in this book focuses on how to design (particularly, the procedures for selection of units) and estimation of parameters in a sample survey design. Part I of the book discusses various sample survey designs and related issues.

Understanding dispersion or variations in a characteristic is basically to know why such variations occur. For example, why some people in an area suffer from a disease while others do not, why some people prefer to use a commodity or service while others do not, why people have varying number of children or marry at different ages and so on. This essentially means examining the role of other characteristics that could be related to the study characteristic and thus explaining the variations in it. The study characteristic is often called the 'dependent' or 'output' variable and the other characteristics that are believed to be related to the dependent variable are called 'independent' or 'input' variables.

Examination of associations or correlations between variables can facilitate comprehending the relationship between variables and hence how one influences the other. Higher the association between an independent variable and a dependent variable, greater is the likelihood that it affects the dependent variable and therefore could help in understanding its variations. However, simple associations are not always enough to reveal the "true" or "causal" relationships between variables, particularly in social science since social phenomena are rarely so simple as to be explained by a single factor (Freedman, 1999). The other factors that obfuscate the measurement of causal relationship between the dependent variable and the study variable (independent variable) are known as 'covariates'. Their presence can hamper the measurement of the causal relationship between a study variable on the desired output. It is thus important to understand particular designs that can handle such problems. Such designs are varyingly termed as experimental or evaluation studies and in part II, this book discusses various designs to generate necessary data to facilitate obtaining a causal effect. In this book, we term this design as evaluation design.

The two designs – sample survey design and evaluation design – can help each other in attaining their objectives. Use of evaluation designs, particularly the recent emphasis on evaluating public programs, can be strengthened using appropriate survey designs. It is often necessary to control large number of covariates to isolate a program effect. One viable alternative is to have samples of both treatment and control group selected randomly from a population. Knowledge of survey design will be an added advantage in such cases. Measurement of program effect requires an estimate of its sampling variance. Familiarity with survey design can facilitate this. As surveys become popular, it will be pertinent to ask whether the survey data can be utilized more profitably to understand a causal relationship. The survey data can provide additional assurance about the external validity. Therefore, if a survey can provide the required data to obtain a treatment effect, should it not be preferred? For example, a large scale survey on health might provide the required information to study the effect of smoking on health.

Reducing bias, estimation of errors, and addressing external validity are common concerns in both the designs. Understanding of both, sample surveys and evaluation designs, can go a long way in successfully conducting fruitful researches. The present book makes a humble attempt to emphasize the role of the two designs to promote scientific research. It provides basic knowledge of the various designs to facilitate their applications. In this process, it is necessary to utilize a few propositions that are the basis of measuring an underlying concept. Mention may be made of few concepts such as 'sampling variance', 'bias', and 'error' of an estimate. The emphasis here is to comprehend a concept and know how to measure and interpret it. The details of derivation of a concept have been avoided. Although such theories are essential for the development of a discipline, their avoidance here is mainly to support applications of the designs.

The book is divided into two parts and ten chapters. In part I (Chapters 1–5) we discuss survey design covering several concepts. In Chapter 1, we discuss probability and non-probability sampling designs, concepts of bias and error, and a guideline to choosing desirable design. Chapter 2 discusses simple random sampling, stratification, stratified random sampling, method of and gain in stratification, probability proportional to size (PPS) sampling, cluster sampling and estimation of mean and proportion and their variances in different designs. It also includes an important discussion on role of weights and its calculation. Chapter 3 deals with multistage design, cluster sampling, effect of clustering and provides several useful guidelines in achieving desired design. In Chapter 4, we consider the application of probability sampling in the presence of defective frames or when no frame is available. It also talks about household listing, its importance and alternatives to household listing. We bring in detailed discussions on non-sampling error in Chapter 5.

In part II (Chapters 6–9) we discuss evaluation design. In Chapter 6 we cover types of evaluation designs including a discussion on bias and error. We discuss designs for measuring cause and effect relationship and related estimation issues in Chapter 7. Chapter 8 deals with issues of allocation in order to achieve balance between treatment arms in an evaluation. Chapter 9 discusses, in brief, various statistical tests for measuring impact.

Chapter 10 looks into the practical application of both the designs.

# Acknowledgments

This body of work could never be completed without inspiration and support from several individuals who not only motivated the authors to pen down their practical experiences with sampling design, but often also challenged them intellectually thereby enriching the contents of the book. The authors would like to express their gratitude to some of these individuals:

The book would have been impossible without intellectual stimulants from Professor Faujdar Ram, Director and Senior Professor, International Institute for Population Sciences (IIPS), Mumbai, Mr S. K. Das, Ex-Director General, Central Statistical Organisation (CSO), India, Dr Fred Arnold, Senior Fellow, ICF International, Rockville, Maryland, USA, Dr Arvind Pandey, Director, National Institute for Medical Statistics (NIMS), New Delhi, and Dr Shireen Jeejebhoy, Senior Associate, Population Council, New Delhi. The authors are also grateful for the technical support provided by Dr Abhishek Singh, Dr Aparajita Chattopadhyay, Dr Laxmikant Dwivedi, and Kaushalendra Kumar - all from the faculty of IIPS, Mumbai. D. D. Mestri, Library and Information Officer, IIPS, Mumbai helped in multiple ways during the course of drafting the book. Our Special thanks also go to Dr Anupendu Gupta, Ex Deputy Director General, Geological Survey of India for motivating us to write the book in the first place.

The authors are also grateful to a number of organizations for their support before and during the writing of the book. Special mention needs to be made of organizations such as the International Institute of Population Sciences (IIPS), Mumbai, the CDC Foundation, Atlanta, the National Council of Applied Economic Research, New Delhi, the Population Council, New Delhi and Economic Information Technology, Kolkata where the authors developed the necessary skills and expertise for writing this book.

We are also indebted to the unknown reviewers who had devoted their precious time to go through the manuscript and give their valuable comments and suggestions. We also thank all the authors whose works we have referred to in the book.

We will be failing in our duty if we do not appreciate the efforts of Manish Choudhary at Cambridge University Press, New Delhi for seeing us through this book. Thank you Manish Choudhary, for your constant inspiration and support and for being patient with us when we missed several deadlines during the submission and evaluation process of the book.

Special thanks go to Jayati Roy, Madhumita Das and Rita Roy, wives of the authors, without whose supportive roles this book would not have seen the light of the day.

We hope that the book provides sufficient knowledge and stimulants to the readers to develop their own statistical designs for applications in real life situations. It would be fulfilling to see this book being equally helpful to users with or without knowledge of statistics, and users from the development sector as well as academics.

Chapter 1

# Introduction to Sample Survey Designs

## 1.1 INTRODUCTION

The objective of a sample survey is to make inferences about a population parameter by observing a portion or sample from it. It is natural to expect that this phenomenon of interpreting about a wider group (a population) on the basis of a sample from it will have some margin of error in the result. How the samples are drawn or selected, while designing a sample survey, is important to ensure that the inference drawn is both valid and reliable. A proper design not only helps obtain a valid idea about the wider population parameters, but also provides a margin of error in an estimate. In addition, the theory can also guide in choosing an alternative design so that the margin of error can be minimized. Different designs that can be employed in a sample survey are discussed in the subsequent chapters. This chapter elaborates few terminologies that will facilitate further discussion on the specific methodologies.

## 1.2 POPULATION, UNITS AND SAMPLING UNITS

Aggregates of units or elements comprise a population. A study unit is a unit or a member of a study population, as in the case of a study concerning human beings, it can be individual persons, families or households in a given geographical area. A study unit, however, can be any living or non-living subject in a study. For example, if the interest is to estimate the number of fruits in an orchard, the study units will be the fruits in the area. Further, if the interest is to estimate the number of words in a book, words of the book will form the study unit.

A distinction, however, needs to be made between a study unit and a sampling unit. Considering an example of sampling from a human population, it is often preferred to select households before selection of individuals – the study units in a population. In this case, the households, consisting of individuals, which facilitate the process of the sample selection are called sampling units. However, if our aim is also to estimate household income, then household will be both the study unit and the sampling unit. Therefore, a sampling unit can also be the study unit in a study. Generally, sampling units are a higher level or a group of units. Their purpose is to help selection of the study units. While selecting a sample, as will be discussed later, one may resort to selection at different stages. For example, for selecting households in rural areas of a country like India with state, district, block and village as the administrative subdivisions, one may consider selection at different stages. It is possible to select a few districts at the first stage, few villages from the selected districts at the second stage, and finally few households from the selected villages at the third stage. The sampling units at each stage of this selection are districts at the first stage, villages at the second stage and households at the last stage. Thus, a study unit, which is household in the present case, is unique referring to the ultimate elements or units and is the focus of study, whereas sampling units can vary depending upon the units (generally areal units) that are selected at a particular stage of sampling.

It is also necessary to distinguish between a *study* population, a *target* population and an *external* population. For example, to understand teenage pregnancy in an area, all girls aged 13–19 years residing in the area becomes the target population. However, while implementing the survey design, more often than not, only households that might have girls aged 13–19 years residing in it are considered for selection. Hence, households in which girls aged 13–19 years reside become the study population and it excludes all those staying in institutions like hostels, boarding schools, etc. The difference between a study population and a target population, however, may or may not be there in a particular case.

Sometimes, for sake of convenience or to make a design more efficient in terms of cost, the target population may be very different from the study population. For example, suppose the interest of a study is to estimate the percentage of obese population among the adult population in a city and suppose there exists a big park in the city where all the people come for the purpose of jogging. One way to estimate the obese population could be to draw a sample from among those who visit the park, which would be both economical and convenient. The study population, in this case, is the population that visits the park can be apparently different from the target population, the population of the city. The two populations can be regarded as similar for the purpose of the estimation, if all the factors that influence obesity are similar in the two populations. In other words, if the chances of visiting the park are similar among both obese and non-obese, the two populations could be regarded as similar for the present purpose. Otherwise, if generally the obese are more likely to jog and hence visit the park, the estimate of obesity on the basis of the study population is likely to be positively biased (i.e., overestimated).

The external population, in this case, could be the population of the entire district in which the city is situated or even the country's population. Whether or not the estimate derived from the study population (from which the sample is selected) will hold good or if it can be generalized to the external population will depend on the different factors that influences obesity (the parameter in question). If the study population and the external population are similar in terms of these factors, a generalization could be made.

Study population, target population and external population can be same or different from each other. A probability sample drawn from a study population will provide a valid estimate of its parameter. But whether it would be so for the target population and external population, if different from the study population, would need further assessment.

To describe a population, there is also a need to have a time reference, particularly if the parameter under study changes with time. In the above example, on measuring obesity in a population, one needs to fix a time reference, such as in year $t$ or during $t$ to $t + k$. It is imperative to have the survey period (the time period when obesity measures are actually collected from a sample) to coincide with the reference period.

## 1.3 SAMPLING DESIGN

A sampling design is the process to accomplish the following tasks:

- number of units to have in a sample (sample size),
- the procedure of selection of the required units (how to select the required units into the sample),
- the procedure of estimation of the required parameters (to transfer the information collected from the sample into meaningful measure to gauge the population parameter), and
- provide an idea about the likely error that one would commit in making the inference from the sample to the population.

There are different sampling techniques that can be employed to draw a sample. These techniques can be broadly classified into two types: a) probability sampling and b) purposive sampling.

## 1.4 PROBABILITY AND PURPOSIVE SAMPLING

### 1.4.1 Probability sampling

In a probability sampling, each and every unit in the study population is given a known and non-zero chance of being included in a sample. Although each and every unit in a population has a chance of getting selected into a sample, the actual selection of a unit, which is carried out mechanically (discussed later), is a chance event and depends on its probability of selection. Hence, the group of units that will ultimately form a sample is not known in advance and there can be a variety of options or alternative samples, once a design, in terms of sample size and the chance of selection of each unit are specified. This phenomenon of exercising the restriction of a known and non-zero probability of selection for each unit into a sample led to the development of sampling theory to provide a basis not only for arriving at a valid estimation procedure but also to have an idea about the error in estimation. Probability sampling has several applications and, in fact, has been applied extensively – so much so that the application of purposive or non-probability sampling is becoming a rare phenomenon. The present book deals only with the sampling methods that use the probability sampling techniques.

Although the definition given for a probability sampling is simple, the chance of violating it, particularly at the time of sample selection, is high. For example, a violation in providing a chance of selection to each unit can occur in many ways, some of which are very subtle and occur unknowingly. Similarly, one needs to take proper care to ensure that there is no deviation from the proposed scheme of the probability of selection given to each unit. These issues are discussed in Chapters 5 and 10.

### 1.4.2 Purposive Sampling

Purposive sampling is done based on subjective selection of a sample, largely based on a few assumptions. One assumption is that there is homogeneity within a population under study. In other words, the variation in a variable that is being studied is practically negligible. For example, as we all know, few grains of rice (or in other words taking a sample of) from a large bowl are enough to check whether it is cooked properly.

Estimate of population size, in the earlier period, is determined by using this principle. Estimating certain rates or ratios, such as population density (number of population per square kilometres) or ratio of number of school-age children to adult population, or number of civilians per 1000 military population, one could estimate the total population. For example, if total area ($X$) is known, one could estimate the density of population ($d$) by counting population in few selected areas as

$$d = \frac{y}{x}$$

where $y$ is the total population size observed in area $x$. Hence, using simple extrapolation method, the estimated total population size in area $X$ will be

$$\text{Total population} = X \cdot d = X \cdot \frac{y}{x} \tag{1.1}$$

In other words, for the estimate of population of $X$, extrapolation is done on the basis of that observed in the sample (for area $x$). For such generalization or extrapolation to be valid, assumption of homogeneity of the distribution should be true for the entire area $X$. If not true, the extrapolation will be invalid, there will be error in the estimate and its magnitude will depend on how the density of population varies in the area $X$. It is, however, not possible to provide any idea about the extent of the error, if the sample is drawn purposively.

Another assumption in the purposive sampling is that the distribution of units in a population is random. If population units are distributed randomly, selection of units from any section of such a population would provide a random sample. Simply stated, this means that characteristics of two units placed adjacent to each other are as similar or dissimilar as any two units selected randomly from the population. For example, in a village, socio-economic conditions of households adjacent to each other are more likely to be similar in terms of certain characteristics. This phenomenon is known as 'clustering' and will be discussed later (Chapters 2 and 3).

In recent years, a type of purposive sampling, known as 'quota sampling' has gained some popularity. In this method, the assumption of homogeneity is relaxed.

Let us discuss this by considering the estimation of total population in area X, discussed above. We may say that the density of population will not be homogeneous throughout but will vary by, say, type of land: a) non-agricultural land, b) agricultural land, and c) fallow land. Let the total area of these three are $X_1$, $X_2$, and $X_3$, respectively, such that $X_1 + X_2 + X_3 = X$. One then needs to subdivide the sample $x$ and select from each of the three categories. Let these be $x_1$, $x_2$, and $x_3$, respectively, which are called quota from each category. Assuming homogeneity within each category and using equation 1.1.

$$\text{Estimated total population} = X_1 \cdot \frac{y_1}{x_1} + X_2 \cdot \frac{y_2}{x_2} + X_3 \cdot \frac{y_3}{x_3} \tag{1.2}$$

where $y_i$ is the estimated population in area $x_i$.

It can be noticed that, in this case, the extrapolation is done within each category and then added up. For this extrapolation, the required assumption would be that the population is homogeneously distributed within each category of land. This is more tenable than the assumption made earlier for the entire population. As for another example, in an opinion poll survey, suppose it is decided that education, occupation and gender are the three variables that are important in understanding the voting behaviour. One can have different categories with the help of these three variables (for further discussion on creation of categories, see Chapter 2). Interviewers can then be given quotas ($x_i$), number of persons to be interviewed in a category and obtain each person's preference for an individual political party. If $y_{ij}$ denotes the number of persons in $i$th category who favoured the $j$th party, extending equation 1.2, it would be possible to estimate the overall proportion of the population who favours $j$th political party. However, in many situations, it is not possible to correctly assess the different variables that should be considered to capture the variations in the study variable. Population size in each category ($X_i$'s), which are required for the estimation (see equation 1.2), may also not be available and the assumption of perfect homogeneity within a category may be difficult to ensure. A cost-effective approach is often to adopt a probability sampling design.

## 1.5 FRAME

A frame is the basis for drawing sampling units in a probability sampling. If sampling units are elements, a frame is supposed to provide the list (detailed address to facilitate contacting a selected unit) of all the elements. A frame, if perfect, provides the list of all the units occurring once and only once and should not include any other elements. That is, if there are $N$ units in a population, the frame should have only $N$ entries in the list giving contact details of each of the units.

If sampling units refer to some macro-level units such as villages and districts, such area units together should cover the entire population under study. Suppose villages are the sampling units. Each village should have a clearly defined boundary; non-overlapping to each other and together they should comprise the total population. It may be noted that the contact details of persons or households within each village is not needed. Such details would be required only for the villages that actually get selected. Auxiliary information on units or sampling units, if available, can (as will be discussed later) strengthen a design. A frame, alternatively called a sampling frame, provides the basis for implementing a probability design.

## 1.6 BIAS AND ERROR

When a design is specified, as mentioned, there may be several alternative samples that could be selected. Each possible sample will have an error that is defined as the deviation between a sample estimate and the corresponding population parameter. Assuming that the interest is to estimate the population mean, such error would be equal to the difference between a sample mean and the population mean. For a given design, one can have a large number of errors, each corresponding to an alternative sample value. Therefore, it is possible to have one value that will provide a summary measure of error (taking differences from the alternative samples) for a design.

Now the questions are why do we require a summary measure of error for a design and how do we obtain it when, in practice, only one sample value is available.

We require a summary measure of error because as alternative samples are possible, in a probability sampling design, for the actual selection, it is necessary to provide 'protection' against all the possible samples of a design. A summary measure of error can help us provide such an idea for the design.

While it is true that, in practice, only one sample gets selected by applying the design and will be available for the estimation. The sampling theory, however, helps get an estimate of the error for a design from a single sample. The summary measure of error is denoted by mean square error (MSE). The MSE comprises of two components: bias and error.

It needs to be emphasized that the deviation (between a sample estimate and the population parameter) caused by the sample selection is generally referred to as the error due to sampling or simply by 'sampling error'. The bias and error can also occur at several other stages in the entire process of a survey design. For example, a deviation between a sample estimate and the population parameter can happen at the stages of data collection, data entry and its processing. These latter types of causes for the deviations are known as the non-sampling errors (discussed in Chapter 5). Non-sampling errors occur during information generation and they are common to both sample survey and census. Although MSE refers to both the sampling and non-sampling errors, our discussion here only confines to the sampling error.

There is, however, a subtle difference between the terms 'bias' and 'error', both of which tend to increase the MSE. Let us discuss them in the context of the sample mean.

If $\bar{y}_i$ denotes the sample mean in the case of the $i$th alternative sample with $\bar{y}$ as the mean of all the sample means in a design, then

$$\text{MSE}\ (\bar{y}) = \frac{\sum_{i=1}^{K}(\bar{y}_i - \bar{Y})^2}{K} \tag{1.3}$$

where $\bar{Y}$ is the population mean and $K$ is the number of alternative samples possible in the design.

It basically takes into account how each sample mean deviates from the overall population mean. Denoting $E(\bar{y}_i)$ as the expected value or mean of all the possible sample means, the above equation can be written as

$$\text{MSE}\ (\bar{y}) = \frac{\sum_{i=1}^{K}\left((\bar{y}_i - E(\bar{y}_i)) + (E(\bar{y}_i) - \bar{Y})\right)^2}{K}$$

since the product term is zero, it can be written as

$$\begin{aligned} &= \frac{\sum_{1}^{K}\left[\left(\bar{y}_i - E(\bar{y}_i)\right)^2 + \left(E(\bar{y}_i) - \bar{Y}\right)^2\right]}{K} \\ &= \left(\bar{y} - \bar{Y}\right)^2 + \frac{\sum_{i=1}^{K}\left(\bar{y}_i - \bar{y}\right)^2}{K} \end{aligned} \tag{1.4}$$

The first component on the right-hand side of equation 1.4 indicates square of the deviations between mean of all possible sample means in the design from the population parameter and is termed as the bias occurring in the sample estimate due to the sampling design.

Why is it called a bias and not an error? What does the presence of bias in a design signify?

While there are $K$ alternative samples possible in a design, the focus is on their mean value ($\bar{y}$). A desirable property of a design is that $\bar{y}$ should equal the population mean ($\bar{Y}$). In other words, while some of the values of $\bar{y}_i$ will be higher than $\bar{Y}$ and the deviations will be positive, for the others the deviation will be negative and $\bar{y}$ would equal $\bar{Y}$ if the positive and the negative deviations cancel out. This means that the sample mean of the design will be regarded as an unbiased estimate of the corresponding population mean. That is, the first component of equation 1.4 would be zero. It is then that whichever sample actually gets selected, the sample mean of it will be regarded as a valid estimate of the population mean.

If there is a bias, that is, the positive and the negative deviations do not cancel out, it will mean that $\bar{y}$ is either higher or lower than the population mean, $\bar{Y}$. In such a case, the sample mean cannot be regarded as a valid estimate of the corresponding population mean. Hence, one needs to have an estimate of the bias and make appropriate corrections to get a valid estimate.

To clarify further, consider the case of estimation of income in a population consisting of rich as well as poor in equal proportion. That is, half the population is considered rich and the other half as poor. If a sample design (with a fixed sample size $n$) provides greater chances of selection of poor people in the sample, it is quite natural to expect that it will end up having poor people in greater proportion into a sample compared to that existed in the population, and hence, the mean of all possible sample means is expected to be lower than the population mean or that it will be negatively biased. Similarly, if the design assigns greater chances of selection to the rich compared with the poor, the mean of all possible sample means would be higher or positively biased.

The sampling theory shows that the mean of all possible sample means would be equal to the population mean or unbiased only when the design is equal probability of selection method (EPSEM); that is, it assigns equal chance of selection to each and every unit to be selected in a sample.

The second component of equation 1.4 gives an idea about the extent to which alternative sample means, in a design, vary from their mean, and is known as the sampling variance of mean and its square root as the standard error of the mean. Hence,

$$\text{MSE}\ (\bar{y}) = B^2 + \text{sampling variance of the mean} \tag{1.5}$$

where $B$ is the extent of the bias.

The square root of MSE is known as the standard error of the mean and for an unbiased estimate, MSE is equal to the sampling variance of the mean.

Following is an illustration to better understand the meaning of both the bias and the error by considering a small hypothetical population.

### Example 1.1

The daily wage of a population consisting of six persons is as follows:

The small population under consideration has six members A–F, whose daily wage is given below.

| Person | A | B | C | D | E | F |
|---|---|---|---|---|---|---|
| Daily wage | 50 | 60 | 1200 | 100 | 1400 | 70 |

For this small population, the MSE of sample mean is computed for a design where a sample of three individuals are selected one by one without replacement (i.e., once a unit is selected, it will not have another chance of selection) giving equal chance of selection to each units.

$$\text{The population mean } (\bar{Y}) = \frac{(50+60+\cdots+70)}{6} = \frac{2880}{6} = 480$$

With this sample design, there are 20 $\left(\frac{6!}{3!3!}\right)$ different samples of size 3 that can be chosen. Table 1.1 provides the necessary computations to understand the components of bias and error in the sample mean.

Few characteristics of the population and the distribution of the sample deviations from population mean may be worth noting. Out of the 6 units, 2, C and E, are rich whereas the

TABLE 1.1 Estimates of average daily wage and deviation between sample mean and the population mean for the 20 alternative samples drawn from the population shown in example 1.1

| S. No. of samples | Units in a sample | Sample mean ($\bar{y}$) | Deviation ($\bar{y}-\bar{Y}$) |
|---|---|---|---|
| 1 | ABC | 436.67 | –43.33 |
| 2 | ABD | 70.00 | –410.00 |
| 3 | ABE | 503.33 | 23.33 |
| 4 | ABF | 60.00 | –420.00 |
| 5 | ACD | 450.00 | –30.00 |
| 6 | ACE | 883.33 | 403.33 |
| 7 | ACF | 440.00 | –40.00 |
| 8 | ADE | 516.67 | 36.67 |
| 9 | ADF | 73.33 | –406.67 |
| 10 | AEF | 506.67 | 26.67 |
| 11 | BCD | 453.33 | –26.67 |
| 12 | BCE | 886.67 | 406.67 |
| 13 | BCF | 443.33 | –36.67 |
| 14 | BDE | 520.00 | 40.00 |
| 15 | BDF | 76.67 | –403.33 |
| 16 | BEF | 510.00 | 30.00 |
| 17 | CDE | 900.00 | 420.00 |
| 18 | CDF | 456.67 | –23.33 |
| 19 | CEF | 890.00 | 410.00 |
| 20 | DEF | 523.33 | 43.33 |
| Sum of deviation | | | 0.0 |

remaining 4 are relatively poor. The alternative samples, in the design with sample size 3, such as ACE, BCE, CDE and CEF where both the rich individuals get selected, shows a large average wage and consequently their deviations from the population mean are positive and large. Similarly, there are four alternatives, ABD, ABF, ADF and BDF where the samples consist of all the three belonging to the poor. This means that their sample means are way below the population mean. In the remaining 12 samples, the deviations are not too much as they consist of 2 from the poor and 1 from the rich individuals. However, since the design is EPSEM, that is, it gives equal chance to each individual to be selected in the sample, the average of all possible samples is equal to that of the population mean (sum of the deviations of all the possible sample means from that of the population mean is zero or zero bias). Thus, it gives an unbiased estimate (the proof of this theory can be obtained from any text book on survey sampling). Alternatively, we say that the expected value of sample means ($\overline{y}$) would equal to the population mean ($\overline{Y}$). Hence, the sample mean estimates the population mean without any bias, if the procedure of selection provides equal chance to each and every element in a population to be included in a sample. It is clear from Table 1.1 that none of the alternative sample means is exactly equal to the population mean. So, the property of being unbiased does not in any way guarantee that the sample mean from the sample that actually gets selected by such a design would equal to the population parameter. It would be pertinent to obtain the value of MSE of sample mean.

Since the estimate is unbiased, the first component (equation 1.4) is zero, it will be

$$\text{MSE}\ (\overline{y}) = \frac{(-43.33)^2 + (-410.00)^2 + \cdots + (43.33)^2}{20} = 67953.33$$

Note that this will be equal to its sampling variance as the mean is an unbiased estimate of the population mean. Hence,

$$\text{SE}\ (\overline{y}) = \sqrt{\text{MSE}(\overline{y})} = 260.68$$

where SE ($\overline{y}$) is the standard error of mean in the design.

While the implication of SE ($\overline{y}$) is discussed in Chapter 2, it may be noted here that it is an error and denotes the likely deviation between a sample and the population mean. It can be noted that its magnitude is quite large.

Now, consider another design in which we draw a sample of three units from the six individuals mentioned above. Suppose one decides to select C and E with certainty (probability of one) and one of the remaining four with equal probability. The four possible samples and the corresponding sample means are shown in Table 1.2.

TABLE 1.2 Estimates of average daily wage and the deviation of sample mean from the population mean for the four alternative samples from the second design (one selected randomly from A, B, D and F and both C and E are included with certainty, see data in example 1.1).

| S. No. sample | Units in sample | Sample mean | Deviation from population mean ($\overline{Y}$) |
|---|---|---|---|
| 1 | ACE | 883.33 | 403.33 |
| 2 | BCE | 886.67 | 406.67 |
| 3 | CDE | 900.00 | 420.00 |
| 4 | CEF | 890.00 | 410.00 |

It can be seen that the deviations are not randomly distributed around the population mean, as in the previous design. The four alternative samples are all on the higher side. In this case,

$$\bar{y} = \frac{883.33 + \cdots + 890.00}{4} = 890.00$$

$$B = (890.00 - 480.00) = 410.00$$

It can be noticed that although the extent of bias is large, the alternative sample means are quite close to each other, and the variance from their mean is quite small, equal to 38.9. Therefore, applying equation 1.4 we get,

$$\text{MSE}(\bar{y}) = 410.00^2 + \frac{(883.33 - 890.00)^2 + \cdots + (890.00 - 890.00)^2}{4}$$

$$= 16813.89$$

$$\text{SE}\ (\bar{y}) = \sqrt{\text{MSE}(\bar{y})} = 410.05$$

This example is discussed further in Chapter 2, where it will be shown as to how to reduce the error and still have an unbiased estimate.

## 1.7 FEW GUIDELINES FOR A DESIRABLE SAMPLING DESIGN

Although a probability sampling is not a must, it is a desirable choice, particularly for undertaking a scientific investigation. Purposive selection can lead to unknown biases that can be serious. Following points are worth noting.

- A probability sampling design with known and non-zero probability of selection to each unit of a population for inclusion into a sample is desirable. It provides an idea about the bias, if any, as well as error in an estimate.
- As far as possible, an EPSEM design should be considered. It helps in obtaining an unbiased estimate of a population parameter. For example, a sample mean can then be regarded as equal to the population mean. Sampling theory not only facilitates estimation of the magnitude of the error, but also selection of an appropriate design to minimize it.
- There are situations where it is neither possible nor necessary to have an EPSEM design. In such situations, bias occurring due to unequal probability of selection of units can be compensated by using appropriate sample weights (discussed in Chapter 2, also see Potthoff et al., 2008). If designed properly, standard error of an estimate can also be obtained.
- Error is generally less of a worry compared to bias in a design.
- It needs to be emphasized that the bias in an estimate of sample mean does not decline with an increase in the sample size, whereas its standard error can be reduced by increasing the size. It is, therefore, desirable, particularly in large-scale sample surveys, to keep the design as close to EPSEM as possible, unless there are genuine reasons to believe that deviating from EPSEM would help in reducing MSE.
- A sample design should endeavour to capture the heterogeneity or variations in the study variable. For example, if the study variable is economic condition of households

in a population, the design should be such that a sample drawn should consist of households from all the different economic strata in the population. Information on factors like education, place of residence and occupation that are likely to be related to the economic condition, if available, can be fruitfully utilized to achieve this.

Chapters 2–4 of this book attempt to shed further light on the above issues under various situations. Chapter 5 discusses the issue of non-sampling errors.

# Chapter 2

# Basic Sampling Designs

## 2.1 INTRODUCTION

In this chapter, we discuss four basic sampling designs, namely, simple random sampling (SRS), stratified sampling, systematic sampling and cluster sampling. In the process of sample selection, it is possible to select elements directly, if a suitable sampling frame for their selection is available. Often, sampling frame is available only for a group of individual elements, instead of each element. These groups of individual elements, generally some areal units, are commonly referred to as clusters. If the sampling frame for each cluster in a population is available, the sample can be selected either in one stage or at multiple stages. In one stage, in a selected cluster, all the elements comprising it will be automatically included in a sample. Selection in two or more stages, discussed in Chapters 3 and 4, involves selection of elements from selected clusters.

Of the four designs, mentioned above, SRS and systematic sampling are the two basic methods for selection of sampling units (either units or clusters). Both provide equal chance to each sampling unit for inclusion in a sample. Another variant, known as PPS sampling wherein sampling units are selected with varying probability, is also elaborated here. Further, we discuss stratified sampling which is a technique of grouping sampling units in a population before selection where grouping is done to put restrictions on selection of a combination of sampling units to benefit the procedure of selection. The cluster sampling in which all the elements in a selected cluster are included in a sample is generally avoided. However, it facilitates understanding multistage designs. In each of the next four sections, we outline the essential steps in a design. They are as follows:

- Methods of selection of sampling units,
- Estimation of a parameter, and
- Providing estimate of the error such as sampling variance of an estimate.

## 2.2 SIMPLE RANDOM SAMPLING

### 2.2.1 Description

In this design, each and every sampling unit in a population receives equal chance of being included in a sample. The method is discussed assuming individual units as the sampling units, but remains true even if group of units, instead of individual units, is considered. Let there be $N$ units in a population from which $n$ are to be selected for inclusion in a sample. The method of selection adopted is such that every unit, in a sample of $n$ units, has a chance of selection equal to:

$$\text{Probability of selection of a unit} = \frac{n}{N}$$

There are other designs including systematic sampling wherein units are selected assigning equal probability to each of them. Such designs are also referred to as EPSEM.

The distinguishing feature of SRS is that, in addition to giving equal chance of occurrence to each unit, it also assigns all possible samples of size $n$ an equal chance of selection. Consider the case of example 1.1, selecting 3 units with equal probability from the 6 yields 20 possible samples (shown in Table 1.1). Not only that each unit has equal chance $\left(=\frac{3}{6}\right)$ of selection but also the probability of selecting a sample, if SRS design is used, will be equal to $\frac{1}{20}$. For example, let us calculate the probability of selecting a sample consisting of units BCF, assuming that three draws are made one by one and a unit once selected in a draw is not replaced, that is, will not get another chance to occur in the sample. To get BCF, B may be selected in the first draw (with a probability of $\frac{1}{6}$), C in the second draw (with a probability of $\frac{1}{5}$) and F in the third (with probability $\frac{1}{4}$). So, probability of selecting BCF in that order is

$$\frac{1}{6}\frac{1}{5}\frac{1}{4}$$

However, to get BCF order of selection is not important, it can be BFC and so on in any of the 6 (=3!) combinations. Therefore, the probability of selecting BCF is

$$\frac{6}{6.5.4} = \frac{1}{20}$$

This can be easily generalized for selecting $n$ units from $N$ units as

$$\frac{n!}{N(N-1)(N-1)\ldots(N-n+1)} = \frac{1}{Nc_n}$$

Hence, in SRS, all possible samples ($=Nc_n$) of size $n$ have equal chance of being included in a sample. As will be discussed subsequently, though giving equal chance to each unit for selection has certain advantages, it is not necessary to give equal chance to all possible samples.

There are two distinct ways to implement or select units in an SRS design. One is known as with replacement where units selected at each draw is replaced before making the next one. In other words, same unit can occur more than once in a sample. In this case also, the probability of selecting a unit is $\frac{n}{N}$ because at each of the $n$ draws, the probability of selection of a unit remains $\frac{1}{N}$.

Selecting a unit more than once does not make much sense. In fact, in most cases, the SRS without replacement is used. We will refer to the latter technique whenever we use the term SRS. For the other, the term SRS with replacement will be used.

### 2.2.2 Methods of selection

The simplest way to draw units with SRS is the lottery method. If there are $N$ units, one can take that many small pieces of paper (of equal size and shape) and identification of each unit should be put on each piece. Suppose if they are numbered 1, 2, ..., $N$, one must have a list that consists of the identification (name and address) for each number. The piece of papers can be folded in a uniform manner and put in a bowl and $n$ pieces can be drawn one by one. It should be noted that each unit must correspond to one and only one piece so that equal chance is given to each unit for selection.

We can use available random number tables that provide multitude of such numbers that can be used for drawing units with equal probability (Fisher and Yates, 1963). To start, it is necessary to locate a specific point, depending on a page, a row and a column, from where to begin for considering random numbers for selection. The other random numbers need to be obtained sequentially thereafter (Hansen et al., 1953).

Suppose a population consists of 30 units from which three units are to be selected. The units can be identified by numbers ranging between 01 and 30.

Let the first few random numbers selected from the random number table are

...32990516687920...

One needs to look for sequence of two digit numbers in the above series. Thus, in the above example, the first selected unit (consisting of two digits) will be corresponding to 05 because both 32 and 99 fall outside the range considered for the selection. The other two selected units are 16 and 20. In this case, all random numbers above 30 are not considered for the selection. The rules of selection can be suitably modified to use them without disturbing the principle of giving equal chance to all the units for inclusion in the sample.

Now, consider all random numbers between 01 and 60. The numbers between 31 and 60 can be used to select a number between 01 and 30 by subtracting 30 from each such number. If the random number is 44, the serial number 14 (=44–30) will get selected. Therefore, if numbers between 01 and 60 are considered, each unit between 01 and 30 gets two chances to appear. This can be extended further by also considering numbers between 61 and 90, but not 91 to 99. According to above rule, if numbers between 01 and 90 are considered for selection, each unit gets three chances to be selected. If one considers up to 99 it will mean that numbers 01 to 09 get 4 chances of selection, whereas 10 to 30 will get only three chances and the selection will be with unequal probability favouring units 01 to 09.

Let us take another example. Suppose $N$ is 347. Now, since 3 digit numbers can vary between 001 and 999, it is possible to give two chances to each number between 001 and 347 for the selection by considering numbers between 001 and 694 (=347 × 2). Numbers above 694 should not be considered for the selection (since 347 × 3 is more than 999).

One can also generate random numbers from computers or calculators. They usually generate numbers less than one. Either one should be specific and generate a random number between 1 and $N$ or generate a number and multiply that by $N$ and round it upward. MS Excel, however, directly gives random numbers between 1 and $N$ using RANDBETWEEN(1, $N$) function.

In the STATA software, one can use runiform() function to generate random number between (0, 1). In the STATA command line, type *help runiform()* for detailed description of how random numbers between $[a, b]$ or $[a, b]$ can be generated, where $a$ and $b$ are two integers.

### 2.2.3 Estimation of Mean, Total and Need for Weights

Once a sample is selected and values of a characteristic under study are obtained, summary measures or estimates are to be prepared to infer about a population parameter. Let, $y_i$ $(i = 1, 2, \ldots, N)$ be the value of a characteristic for the $i$th unit in a population consisting of $N$ units. Two most common parameters of interest are population mean $(\bar{Y})$ and population total $(Y)$, where

$$Y = y_1 + y_2 + \cdots y_N\text{, and}$$
$$\bar{Y} = \frac{Y}{N}$$

The question is whether the sample mean $(\bar{y})$ calculated on the basis of $n$ selected values of $y$ as

$$\bar{y} = \frac{y_1 + y_2 + \cdots + y_n}{n} \tag{2.1}$$

and, sample total $(y)$ obtained as

$$y = y_1 + y_2 + \cdots + y_n \tag{2.2}$$

can be considered as estimators of the population mean and population total, respectively?

One desirable property of an estimator (the guideline for deriving an estimate) is that it should give an unbiased estimate of the population parameter. In Chapter 1, it has been mentioned that a sample mean, in the case of an EPSEM design, provides an unbiased estimate of the population mean. It must be recognised that this property of unbiasedness refers not to a particular sample estimate but to the estimator in a design. Once a design specifying the procedure of selection of units and estimation is adopted, one can have a number of alternative samples and hence alternative values of an estimate. It (the estimator) is unbiased if average of all possible values of an estimate equals the population parameter.

Statistical theories show that the sample mean in the case of SRS is an unbiased estimate of the corresponding population mean (Cochran, 1977). This point has been corroborated with a small example (example 2.1) below. If an estimator is unbiased, a sample estimate, such as sample mean obtained in a particular case can be considered, for all practical purposes, to be equivalent to the corresponding population parameter. If not, that is, if it is biased, either the magnitude of the bias must be known and adjusted accordingly or there should be assurance that its magnitude is small enough to be ignored.

Let us now discuss about the other estimator, sample total. The question is whether the sample total (equation 2.2) in the case of an EPSEM design would provide, like the sample mean, a valid estimate of the population total? This has been discussed in the context of the data shown in example 1.1. The purpose here is to demonstrate that while this is true for the sample mean, it is not that straightforward for the sample total. According to the SRS design, the 20 different possible samples of three units from the six are provided in Table 2.1. It also shows the mean wage and total wage for each of the possible samples.

**TABLE 2.1** The 20 possible samples of size 3 along with sample mean wage and total wage from the population given in example 1.1

| S. No. of samples | Units in sample | Sample mean wages | Sample total wages |
|---|---|---|---|
| 1 | ABC | 436.67 | 1310 |
| 2 | ABD | 70.00 | 210 |
| 3 | ABE | 503.33 | 1510 |
| 4 | ABF | 60.00 | 180 |
| 5 | ACD | 450.00 | 1350 |
| 6 | ACE | 883.33 | 2650 |
| 7 | ACF | 440.00 | 1320 |
| 8 | ADE | 516.67 | 1550 |
| 9 | ADF | 73.33 | 220 |
| 10 | AEF | 506.67 | 1520 |
| 11 | BCD | 453.33 | 1360 |
| 12 | BCE | 886.33 | 2660 |
| 13 | BCF | 443.33 | 1330 |
| 14 | BDE | 520.00 | 1560 |
| 15 | BDF | 76.67 | 230 |
| 16 | BEF | 510.00 | 1530 |
| 17 | CDE | 900.00 | 2700 |

*(Contd)*

TABLE 2.1 (Contd)

| S. No. of samples | Units in sample | Sample mean wages | Sample total wages |
|---|---|---|---|
| 18 | CDF | 456.67 | 1370 |
| 19 | CEF | 890.00 | 2670 |
| 20 | DEF | 523.33 | 1570 |

$$\text{The mean of all sample means} = \frac{436.67 + 70.00 + \cdots + 523.33}{20}$$
$$= 480.00$$

which, as already mentioned, equals the population mean indicating that the sample mean, in this case, is an unbiased estimate of the population mean, that is,

$$E(\bar{y}) = \bar{Y} \tag{2.3}$$

Thus, the mean of all possible sample means, in the design, will equal its population mean. However, this is not true for sample total, as the mean of all possible sample total

$$= \frac{(1310 + 210 + \cdots + 1570}{20}$$
$$= 1440$$

is, however, much less, in fact, half of the actual population total of 2880. Hence, the sample total as such is not an unbiased estimator of the population total.

Why is sample total not an unbiased estimate when the sample mean is? It is worthwhile, in this context, to mention the fundamental theorem which states that if there exist $n$ independent and unbiased estimators of a population parameter, then their mean will also be an unbiased estimator of it (Som, 1996).

The inherent assumption in an equal probability selection design is that the units in the population are homogeneous (of more or less equal value). Hence, each sample unit can be considered as an unbiased estimate of the population mean. In view of the above theorem, therefore, mean of all sample values would be an unbiased estimate of the population mean.

However, even if units are of equal value, what we also require is an estimate of the sum total of $N$ values. The value of a single individual cannot be equal to the total of $N$ values. Imagine that in a store, there is a stack of 200 books each costing approximately Rs. 50. To buy one book, one needs to have Rs. 50, but to have all the 200 it would require Rs. 10,000. In other words, while each $y_i$ cannot equal $Y$, $Ny_i$ can be considered as an unbiased estimator of population total, in an EPSEM design.

Hence, as per the above theorem, their mean from the sample

$$\frac{\sum_i Ny_t}{n} = N \cdot \bar{y} = N\bar{Y} = Y$$

Note that $\bar{y}$ is an unbiased estimate of $\bar{Y}$. That is, if each unit in a sample is multiplied by a factor $\frac{N}{n}$, they can be simply added to get the population total (similarity between this and the discussion in equation 1.1 of Chapter 1 may be noted. The application of SRS provides the basis for the assumption of homogeneity among units in the population). This factor is known as the weight.

Another way to see this is because $n$ units selected with EPSEM have to represent $N$, each unit is to be considered as equivalent to or representing $\frac{N}{n}$ units. This is the logic used in a simple extrapolation done to obtain the total of a characteristic by studying few of them. For example, if 10 schools, selected with equal probability from 60 schools, in an area, yields 500 students, then the total number of students in the area can be estimated as $3000\left(=\frac{60}{10}\times 500\right)$. This will hold good as long as the assumption that the schools are homogeneous in terms of their size (of students) is true. The equal probability selection, which implicitly assumes the homogeneity, enables to obtain such a parity which can be utilized to estimate the population total. Writing,

$$w_i' = \frac{N}{n} = \frac{1}{f}$$

where $f$ is the sampling fraction or, in this case, the probability of selecting a unit (in the above example of students in school, $f =$ probability of selecting a school $= \frac{n}{N} = \frac{10}{60}$). Hence, $w_i'$ which is defined as the weight of $i$th unit, is inverse of the probability of its selection. Working with $w_i'y_i$ instead of $y_i$ (multiplying each variable value with its weight), it can be checked that the estimates of

$$\text{Sample total } = \sum_{i=1}^{n} w_i' y_i = \frac{N}{n}\left(\sum_i y_i\right) = N\bar{y} = N\bar{Y} = Y, \text{ and}$$

$$\text{Sample mean} = \frac{\sum_{i=1}^{n} w_i' y_i}{\sum_{i=1}^{n} w_i} = \frac{\sum_{i=1}^{n} \frac{N}{n} y_i}{\sum_{i=1}^{n} \frac{N}{n}} = \frac{1}{n}\sum_i y_i = \bar{y} = \bar{Y}$$

will be unbiased estimates of the population total and population mean, respectively. It can be easily checked from Table 2.1 that if each sample value is multiplied by its weight, which is $2\left(=\frac{N}{n}=\frac{6}{3}\right)$, mean of all the possible sample total will be equal to the population total.

#### 2.2.3.1 Normalization of weights

If response is available from all the selected units, there will be $n$ units each having a constant weight $\left(w_i' = \frac{N}{n}\right)$, in an EPSEM design. The total sample size will be

$$\sum_{i=1}^{n} w_i' = \sum_{1}^{n} \frac{N}{n} = N$$

It is, however, customary to indicate the sample size used for estimation of a parameter rather than the population size. To avoid this, a convenient way is to normalize the weights so that its sum, instead of giving population size, would become the sample size that has been used in the estimation.

It can be seen, in this context, that multiplying the weights by a scalar factor, $k$, does not influence the estimation of sample mean. For example,

$$\overline{y} = \frac{kw'_1y_1 + kw'_2y_2 + \cdots + kw'_ny_n}{kw'_1 + kw'_2 + \cdots + kw'_n} = \frac{w'_1y_1 + w'_2y_2 + \cdots + w'_ny_n}{w'_1 + w'_2 + \cdots + w'_n}$$

Hence, considering $k$ as

$$k = \frac{n}{\text{sum of weights}}$$

the normalized weight ($w_i$) can be written as

$$w_i = \frac{N}{n}\frac{n}{N} = 1$$

$$w_i = kw'_i = w'_i \frac{n}{\text{sum of weights}} \tag{2.4}$$

The use of the $w_i$ will serve similar purpose as the $w'_i$ except for the fact that sum of $w_i$ will be equal to $n$ (the sample size) instead of $N$ (the population size).

The normalized weight in the case of SRS will be

$$w_i = \frac{N}{n}\frac{n}{N} = 1$$

In other words, in an EPSEM design, since probability of selection of each unit is equal, it is immaterial whether weight is being used or not. Sample mean, in any case, remains an unbiased estimate, and there is an easy alternative for estimating the population total. It can be estimated as

$$\text{Estimated population total} = N\overline{y} \tag{2.5}$$

There are, however, different opinions regarding whether to normalize weights or not. It should be reiterated that, as far as estimation of a parameter such as mean or proportion is concerned, both normalized weights ($w_i$) and weights without the normalization ($w'_i$) provide the same estimate. In large-scale surveys, where means and proportions and not total of a parameter are the major concern, normalized weights are generally preferred. Since it is quite common to analyse and present data in the form of frequency and bivariate distributions, one might need to have an idea about the relevant sample size. Use of normalized weights can readily provide it and the chances of committing any error on this count while applying a relevant statistical test is avoided. We will use, in this book, the notation $w_i$ for the normalized weight and $w'_i$ for weights without the normalization.

As will be discussed subsequently, in a design, if units are not selected with EPSEM, it would become almost imperative to use weights in order to take care of the bias in estimation of even the population mean. Since the probability of selection of a unit is known in a probability sampling, it is prudent to use weights for each unit (which is inverse of its selection probability) selected in a sample. Multiplying the variate values ($y_i$) with their respective weights facilitate obtaining unbiased estimation of a parameter. Similarly, its sampling variance, using the weights, will provide the estimate of MSE.

#### 2.2.3.2 Role of weights

The use of weights, at the time of analysis, can help take care of the bias in an estimate. It can be seen that if $y_i$ is used instead of $w_i'y_i$, equation 2.2 would provide a biased estimate of the population total. The magnitude of bias will be large. Even for the small example shown in Table 2.1, where the sampling fraction is large, 50 per cent of the population units are selected in the sample, the bias is 50 per cent of the population total. There is another noticeable aspect that happens when a bias is present in an estimate. Suppose it is possible to plot the deviation of estimates from the population value for the different alternative samples (shown in Table 2.1). In an ideal situation, when bias is not present, the deviations will be randomly distributed around a value zero, that is, sometimes an estimate will be higher compared to the population value and for another it will be lower and the positives and the negatives will cancel out. In the presence of a bias, the deviations will tend to show a systematic pattern with the values generally lower or higher compared to the population value. This can be noticed from Table 2.1, that the estimates of the total population, when calculated using equation 2.2, without using the weights, will be systematically lower. They are lower compared to the population value. That is, if an estimate is biased, its deviation from the population value cannot be regarded as an error alone, and the assumptions about the distribution of the error generally made for a statistical analysis may not be tenable.

However, the variation in weights results in an increase in the sampling variance, which can be approximated by a factor, known as a multiplicative factor $\text{Meff}_w$. The value of $\text{Meff}_w$ can be obtained (Kish, 1995) as

$$\text{Meff}_w = 1 + (\text{coefficient of variation of weights})^2 \tag{2.6}$$

That is, if $S_v^2$ is the sampling variance of an estimate in the case of a self-weighting design, then

$$S_{v'}^2 = \text{Meff}_w \cdot S_v^2 \tag{2.7}$$

where $S_{v'}^2$ is the sampling variance when the design is not EPSEM.

If a design uses equal probability selection, the weights are equal to 1 and hence the standard deviation or coefficient of variation of weights will be zero. In this case, according to equation 2.6, $S_{v'}^2$ will be equal to $S_v^2$. Larger the value of the standard deviation of weights (i.e., variation in weights), higher will be the value of $\text{Meff}_w$ and hence greater the departure of $S_{v'}^2$ from $S_v^2$. More discussion on this aspect is provided in Chapters 3 and 10.

The coefficient of variation being a ratio of standard deviation and mean of weights, $\text{Meff}_w$ will remain same irrespective of whether weights ($w'$) or normalized weights ($w$) are used.

Apart from mean and total, estimation of few parameters is discussed as follows:

- Proportion
- A measure for a subgroup of population or subclass estimate

### 2.2.4 Estimation of Proportion

A dichotomous variable, which takes two values if a trait is present ($y_i = 1$) or absent ($y_i = 0$), is represented by a proportion ($P$). Suppose the attribute is whether a person is suffering from a particular disease ($y$) in a population, or persons suffering, $y_i = 1$ or those not suffering, $y_i = 0$. The mean number of persons in the population suffering is

$$\text{Population mean} = \frac{\sum_1^N y_i}{N} = \frac{\text{Number of persons suffering in the population}}{N} = P$$

In a sample of size $n$ selected by SRS,

$$\text{Sample mean} = \frac{\sum_1^n y_i}{n} = \frac{\text{Number of persons suffering in the sample}}{n} = p$$

$p$ is also an unbiased estimate of $P$. Total number of persons suffering in the population is estimated by

$$\text{Estimated total number suffering in the population} = Np.$$

If $N$ is not known, then

$$\text{Estimated total number suffering in the population} = \sum_1^n w_i' y_i$$

### 2.2.5 Subclass Estimates

Estimation of a subgroup of population is very common. For example, even if the objective is to study the proportion of a population suffering from a disease, there would be interest in understanding how the prevalence of the disease varies according to certain socio-economic characteristics. That is, one would require estimates of a subgroup, such as the prevalence among poor or among less educated.

Let $k$ units, among $n$ selected, belong to a subgroup. If the sample of size $n$ is selected from $N$ by SRS, one can consider $k$ also as a simple random sample and calculate accordingly its sample mean. Then the sample mean for the subgroup will be

$$(\bar{y}_k) = \frac{1}{k} \sum_1^k y_i \tag{2.8}$$

It may also be of interest to estimate the total number of diseased persons($y$) among the subgroup. If $K$ is the total number of persons in the population of the particular subgroup, then the estimate will be

$$\text{Estimated total diseased persons}(y) \text{ in the subgroup} = K \cdot \bar{y}_k = \frac{K}{k}\sum_1^k$$

To calculate the total, knowledge of $K$, the total persons($y$) belonging to the subgroup in the population is needed. If this is not available, one can use

$$\text{Estimated total diseased persons}(y) \text{ in the subgroup (when } K \text{ is not known)} = \frac{N}{n}\sum_1^k$$

The latter estimate will, however, be less precise (Kish, 1995, p. 42).

### 2.2.6 Sampling Variance of Estimates

In the case of SRS, MSE of an estimate, as given in equation 1.5, will be equal to its sampling variance. Hence, instead of MSE we discuss sampling variance for different types of estimates.

#### 2.2.6.1 Sampling variance of a sample mean

To gauge what is the sampling variance of an estimated parameter, let us consider example 1.1 and illustrate it in the case of sample mean. Table 2.1 provides the estimated sample mean for each of the 20 alternative samples. The mean of the 20 alternative sample means is equal to 480, the population mean.

The sampling variance of sample mean provides an idea about the extent of variations in the alternative sample means from their overall mean, which is equal to the population mean in the case of SRS. It is possible to calculate the sampling variance as

$$\text{Sampling variance of mean} = \frac{(436.67-480)^2+(70.00-480)^2+\cdots+(523.33-480)^2}{20}$$

$$= 67953.33$$

A pertinent question that can be asked at this stage is how to calculate this crucial measure of error in an estimate. In this small example, it was possible to calculate it, but in practice, it will not be possible to have the alternative sample means and their overall mean. Sampling theory helps us in deriving a formula for the sampling variance. In the case of SRS, the sampling variance of mean can be estimated as

$$\text{Sampling variance of mean} = \frac{(N-n)}{N}\frac{S^2}{n} \tag{2.9}$$

where

$$S^2 = \frac{\sum_1^N (y_i-\bar{Y})^2}{N-1} \tag{2.10}$$

Proof of equation 2.9 can be obtained from any text book on sampling. $S^2$ is very close to the variance of units ($y_i$) in the population, for large $N$. The factor $\frac{N-n}{N} = (1-f)$ is known as the finite population correction (fpc). This factor is assumed to be equal to 1 when $f$, the sampling fraction is small.

It can be verified that

$$S^2 = \frac{(50-480)^2 + (60-480)^2 + \cdots + (50-480)^2}{(6-1)} = 407720$$

Therefore,

$$\text{Sampling variance of mean} = \frac{6-3}{6} \cdot \frac{407720}{3} = 67953.33$$

This value agrees exactly with the one calculated earlier on the basis of the alternative samples.

There is one problem, however, with the formula of sampling variance of mean (equation 2.9). It will generally not be possible to calculate $S^2$, because all the $y_i$ values in a population will not be known. It is only the selected units in a sample for which $y_i$ will be available. That is, it would be possible to calculate

$$s^2 = \frac{\sum_1^n (y_i - \overline{y})^2}{(n-1)}$$

The theory shows that

$$E(s^2) = S^2$$

That is, one can use $s^2$ from the sample as equal to $S^2$. Hence, it is possible to estimate the sampling variance of mean as

$$\text{Estimated sampling variance of mean} = \frac{(N-n)}{N}\frac{s^2}{n} \tag{2.11}$$

A related alternative measure of the error is known as standard error, which is

$$\text{Standard error of sample mean} = \sqrt{\frac{(N-n)}{N}\frac{S^2}{n}} \tag{2.12}$$

The estimated standard error is obtained as

$$\text{Estimated standard error of sample mean} = \sqrt{\frac{(N-n)}{N}\frac{s^2}{n}} \tag{2.13}$$

It is useful to note that three terms are generally used to understand the extent of deviation of alternative sample estimates from its population parameter. They are sampling variance, standard error and 'precision.' Lower the error of an estimate, higher is its precision.

#### 2.2.6.2 Sampling variance of estimated population total

The estimate of population total, as mentioned, is

$$\text{Estimated population total} = N\overline{y}$$

Thus, the sampling variance of the estimated total is

$$\text{Sampling variance of estimated total} = N^2 V(\overline{y})$$

where $V(\overline{y})$ is the sampling variance of sample mean $\overline{y}$. Using equation 2.11, it can be written as

$$\begin{aligned}\text{Estimated sampling variance of total} &= N^2 \frac{N-n}{N} \cdot \frac{s^2}{n} \\ &= N(N-n) \cdot \frac{s^2}{n}\end{aligned} \tag{2.14}$$

The standard error of the estimate is given by

$$\text{Estimated standard error of total} = \sqrt{N(N-n) \cdot \frac{s^2}{n}} \tag{2.15}$$

#### 2.2.6.3 Sampling variance of proportion

The sampling variance of $p$, that is, the proportion of individuals possessing an attribute is

$$\text{Variance of } p = \left(\frac{(N-n)}{N}\right) \cdot \frac{PQ}{n} \tag{2.16}$$

where $P$ is the proportion of a population having the attribute, and $Q = 1-P$. Since the population parameter $P$ is mostly not available, the variance of $p$ is estimated as

$$\text{Estimated variance of } p = \left(\frac{(N-n)}{N}\right) \cdot \frac{pq}{n-1} \tag{2.17}$$

#### 2.2.6.4 Sampling variance of subclass estimates

The sample mean of a subclass estimate based on $k$ units, $\overline{y}_k$ is shown in equation 2.8. The sampling variance of $\overline{y}_k$ can also be estimated considering it as an EPSEM design of size $k$ as

$$\text{Estimated sampling variance of } (\overline{y}_k) = \frac{1-f}{k} \cdot s_k^2 \tag{2.18}$$

where $f$ is the sampling fraction for the subclass variable, that is, the proportion of the subclass variable in the population included in the sample $= \frac{k}{K}$. If the number of subclass variable ($K$) is not known, one can substitute the proportion by $\frac{n}{N}$ (Kish, 1995, p. 42). The element variance of the subclass variable $s_k^2$ can be obtained in the usual manner as

$$s_k^2 = \frac{1}{k-1} \sum_1^k (y_i - \overline{y}_k)^2 \tag{2.19}$$

Let us take an example now.

## Example 2.1

From a population of 5000 households, a simple random sample of 50 was selected without replacement. Information on number of members ($y$) and whether the households have agricultural land ($x$, $x = 0$ for no land and $x = 1$ if a household has some land) are available for each selected household and are presented in Table 2.2.

Table 2.2 Number of members in 50 households and whether they have agricultural land.

| S. No. | 1 | 2 | 3 | 4 | 5 | 6 | 7 | 8 | 9 | 10 |
|---|---|---|---|---|---|---|---|---|---|---|
| $y$ | 4 | 5 | 2 | 3 | 6 | 4 | 6 | 5 | 5 | 2 |
| $x$ | 0 | 1 | 0 | 0 | 1 | 1 | 1 | 0 | 1 | 1 |
| **S. No.** | **11** | **12** | **13** | **14** | **15** | **16** | **17** | **18** | **19** | **20** |
| $y$ | 4 | 2 | 4 | 6 | 6 | 6 | 2 | 4 | 7 | 6 |
| $x$ | 0 | 0 | 0 | 0 | 1 | 1 | 0 | 1 | 1 | 1 |
| **S. No.** | **21** | **22** | **23** | **24** | **25** | **26** | **27** | **28** | **29** | **30** |
| $y$ | 7 | 8 | 5 | 4 | 1 | 5 | 5 | 6 | 8 | 12 |
| $x$ | 1 | 1 | 0 | 0 | 0 | 1 | 0 | 0 | 0 | 1 |
| **S. No.** | **31** | **32** | **33** | **34** | **35** | **36** | **37** | **38** | **39** | **40** |
| $y$ | 8 | 5 | 7 | 3 | 4 | 7 | 7 | 4 | 6 | 5 |
| $x$ | 0 | 1 | 1 | 1 | 1 | 0 | 0 | 0 | 1 | 1 |
| **S. No** | **41** | **42** | **43** | **44** | **45** | **46** | **47** | **48** | **49** | **50** |
| $y$ | 6 | 3 | 6 | 5 | 3 | 5 | 6 | 4 | 3 | 5 |
| $x$ | 0 | 0 | 0 | 1 | 1 | 0 | 1 | 1 | 0 | 0 |

Compute estimate of the following parameters: i) mean household size and total number of persons in the population, ii) proportion and total number of households having no land in the population, iii) mean household size among those with no land, and iv) standard error as well as the 95 per cent confidence intervals of all the estimates.

$$\text{Estimated mean number of members in a households } (\bar{y}) = \frac{4+5+\cdots+3+5}{50}$$

$$= \frac{252}{50} = 5.04$$

In other words, the population has on an average 5 persons per household.
Estimated total number of persons in the population = $N\bar{y}$

$$= 5000.5.04 = 25200$$

Estimated sampling variance of mean household size (equation 2.11)

$$\frac{(5000-50)}{5000} \cdot \frac{3.8351}{50} = 0.0759$$

where $s^2 = \dfrac{(4-5.04)^2 + (5-5.04)^2 + \cdots + (3-5.04)^2 + (5-5.04)^2}{(50-1)} = 3.8351$

Estimated standard error of mean household size

$$= \sqrt{0.0759} = 0.2756$$

and the 95 per cent confidence interval of the mean household size as between (5.04 – 1.96 × 0.2756) and (5.04 + 1.96 × 0.2756) or between 4.50 and 5.58.

This means that with 95 per cent certainty, the alternative sample means from the design (i.e., if 50 samples are drawn with EPSEM from the 5000 in the population) will lie in between 4.5 and 5.6 or the population mean will lie between 4.5 and 5.6.

Estimated standard error of total number of persons in the population (using equation 2.15) = 1378, and its 95 per cent confidence intervals are between 22499 and 27901.

Estimated proportion of households with no land ($p$)

$$\frac{25}{50} = 0.5$$

Estimated total number of households with no land

$$= 5000 \times 0.5 = 2500$$

Estimated standard error of the proportion of households with no land (using equation 2.17)

$$= \sqrt{0.0051} = 0.0711$$

and its 95 per cent confidence intervals are 0.3607 and 0.6393.

Estimated standard error of total number of households with no land

$$= \sqrt{N(N-n)pq/(n-1)} = 355$$

and 95 per cent confidence intervals are in between 1804 and 3196.

For the subclass estimate, average household size and total number of households with no land:

average household size among those with no land

$$\frac{4+2+3+\cdots+5+3+5}{25} = 4.6$$

Estimated standard error of average household size among those with no land (from equation 2.18)

$$= \sqrt{\frac{(1-f)s_k^2}{k}}, \text{ where } f = \frac{n}{N} = \frac{50}{5000}$$

as the number of households with no land in the population ($K$) is not available. $k$ is the number of households with no land in the sample, and $s_k^2$ is the element variance. From equation 2.19

$$s_k^2 = \frac{(4-4.6)^2 + (2-4.6)^2 + (3-4.6)^2 + \cdots + (5-4.6)^2 + (3-4.6)^2 + (5-4.6)^2}{(25-1)}$$

$$= 3.58$$

Therefore, the standard error of average household size for those with no land

$$= \sqrt{0.99 \cdot \left(\frac{3.58}{25}\right)} = 0.38$$

and its 95 per cent confidence intervals are between 3.86 and 5.34.

### 2.2.7 Determination of Sample Size

An initial requirement in a survey design is to have an idea about the sample size or the number of units that needs to be selected. While determining sample size, it is necessary to take into account following few related aspects:

**(a)** The size is intimately related to the extent of error in an estimate (equations 2.11 and 2.16). As has been mentioned, the standard error of an estimate is inversely related to the sample size. Hence, larger the size of sample better it is in terms of controlling the error (the error due to sampling).

**(b)** The size, in general, is directly related to cost or the requirement of resources to conduct the survey.

**(c)** The size has a link with the extent of non-sampling error. Too much increase in the size may adversely affect the non-sampling error and hence increase the total error.

**(d)** Larger the sample size, greater is the scope for deriving additional estimates, for example, estimates for studying any regional or socio-economic differentials that could be of importance.

Although increase in sample size is beneficial for controlling the standard error of an estimate and expanding the scope of a study, cost and non-sampling errors impose restrictions on increasing the size. A careful balancing act considering time and resources available for a study is important while planning a survey, particularly a large-scale survey.

Consideration of standard error of a parameter being estimated in a study forms the starting point for deriving sample size. The error is an integral part of a sample survey design and one needs to specify a preliminary idea about the error that would be acceptable.

Let us consider the estimation of proportion ($P$) of a characteristic in a population. In SRS, the sample estimate ($p$) is an unbiased estimate of $P$ with the standard error (ignoring the fpc, which is $(1-f)$)

$$\mathrm{SE}(p)=\sqrt{\frac{pq}{n}}$$

The standard error provides an idea about the extent to which $p$ from different alternative samples of size $n$ will vary. Since it depends on $n$, its specification facilitates determination of the size.

How to guess a value of the error in advance? There are two ways to arrive at such an idea.

### Alternative 1

Instead of a point estimate $p$, the standard error of $p$ can help in obtaining an interval for the estimate. Assuming that the alternative sample estimates follow a normal distribution, it is possible to specify that the population parameter would lie in between an interval

$$\{p+d\} \text{ and } \{p-d\}$$

with a given probability or confidence. Denoting the interval $d$ as a function of SE($p$) and considering a 95% confidence interval, one can write

$$d=1.96\sqrt{\frac{pq}{n}}$$

or

$$n=\frac{1.96^2 pq}{d^2} \tag{2.20}$$

Hence, to estimate $n$, we require values of an estimate of the parameter $p$ and interval $d$ that can be tolerated. A rough idea about $p$ will suffice, which can be obtained from past surveys or censuses in the population or from other population having similar characteristic as the present one. It is also possible to provide an idea about $d$. An example can help in understanding it.

Suppose a program administrator, in an area, requires a survey-based estimate of crude birth rate (CBR) that indicates number of births occurring per thousand population in a year. The information is required to decide how many vaccines would be required in order to protect the new born babies from a certain disease. If CBR is 30, that is, $p - 0.03$, a value of $d$ equal to 4 per thousand would mean that the estimate of CBR can vary in between 26 and 34. It is possible to check the vaccine requirement corresponding to these two values of the CBR to decide whether a value of $d = 0.004$ would be tolerable. The sample size required for the estimation would be

$$n=\frac{1.96\times 0.03\times 0.97}{0.004^2}=6987\sim 7000 \text{ population}$$

If average household size is 5, this means a sample of 1400 households would be required. If the interval is deemed to be little too large, the value of $d$ can be reduced. It can be checked that considering $d$ as 2 per thousand would yield a sample size of around 5600 households. It is possible, therefore, to decide a value of $d$ that would suit the purpose of the study.

### Alternative 2

The coefficient of variation (also known as relative standard error) of the sample mean, that is, the standard error per unit sample mean is better and more stable indicator of the standard error. For example, if variable under study is weight of human being and if it is measured in grams, it will have a much higher standard error in comparison to that if measured in kilogram. The coefficient of variation being ratio of standard error to mean is free of the unit of measurement and will not change with the unit. The coefficient of variation of $p$ is

$$\text{Coefficient of variation of } p = \frac{\text{SE}(p)}{p} = \sqrt{\frac{q}{np}}$$

The sample size can be computed as

$$n = \frac{1}{(\text{CV})^2}\frac{q}{p} \tag{2.21}$$

where CV is the coefficient of variation.

In fact, it is always customary to check the required sample size considering a coefficient of variation of 5 per cent. Hence, if approximate value of $p$ is 0.03,

$$n = \frac{1}{(0.05)^2}\frac{0.97}{0.03} = 12933 \text{ population}$$

This means a sample size of 2587 households would be required.

The two alternatives discussed above would yield same estimate provided there is a parity between the parameters used. For example, in Alternative 1, if $d$ is assumed to be 0.004, the estimate of $n$ is 1400 households. Since,

$$d = 2 \times \text{SE}(p)$$

The SE($p$) is assumed to be 0.002 and CV is equal to $6.7\left(=\frac{0.002}{0.03}\right)$ per cent, which is higher than 5 per cent assumed in the present case. This is the reason for getting slightly higher sample size here compared to that in the earlier.

There are three interrelated concepts: an estimate of a proportion ($p$), its precision and sample size required for the estimation. In Alternative 1, the precision is in terms of an interval that provides 95 per cent confidence interval in the distribution of the alternative estimates (this is generally mentioned in terms of 2 SE). In alternative 2, the precision is in terms of the coefficient of variation of the estimate. As $p$ decreases, for a given precision, the sample size, $n$ increases. Alternatively, for a given $n$, as $p$ decreases, the precision also declines.

## 2.3 STRATIFIED SAMPLING

### 2.3.1 DESCRIPTION

Generally, the assumption that units in a population are of equal value will not be tenable because of the heterogeneity in them. Sample selection in such cases with the SRS is likely to have large errors. It is desirable in such cases to group the units into smaller sections or strata so that units within a stratum are as homogeneous as possible. The technique where specified number of units is selected from each stratum is known as stratified sampling. If units within a stratum are selected independently and using the principle of SRS, it will be known as stratified random sampling.

The stratified sampling can ensure better representation of a population. The representation in terms of sample selection that provides better spread of units to capture heterogeneity in the characteristics being studied.

Example 1.1, discussed in Chapter 1, can help understand the concept of stratification. Let us inspect all possible samples of size three in Table 2.1. The samples such as ABD, ABF, ADF, etc. are such where all the units belong to persons getting low wages (Rs. 100 or less) and none from the two who were getting four-digit wages. It can also be noticed that in each of these cases, the departure of sample mean from the true mean is substantial, which contributes towards increasing the sampling variance of mean in the case of SRS.

Can such occurrences be avoided? Stratified sampling tries to provide an answer. One can put a restriction on the selection if prior information on the variation in wages is available. Suppose it is known that four of the six are getting low and two are getting high wages. The actual wages may not be readily available but a fairly good approximation of it can be, for example, obtained by simply observing their place of residence. Hence, the units can be subdivided into two relatively homogeneous groups or strata. One would have the four getting low wages and the other the two having four-digit wages.

The restriction on selection will be to distribute the total sample and select few units from each stratum. The issue of how many to select from a stratum and how to select them will be discussed in subsequent sections. Suppose, it is decided to select two units from the first stratum and remaining one from the second, and the selections are made by giving equal chances to each available unit within the stratum. The distribution of units in the population is presented in Table 2.3.

TABLE 2.3 Distribution of daily wages of six units (shown in example 1.1) according to strata in the population.

| | Stratum 1 | | | | Stratum 2 | |
|---|---|---|---|---|---|---|
| **Unit** | A | B | D | F | C | E |
| **Daily wage** | 50 | 60 | 100 | 70 | 1200 | 1400 |

The number of possible samples under the present design of selecting two from stratum 1 and one from stratum 2 would reduce to 12. Table 2.4 provides list of all possible samples and their mean.

TABLE 2.4 Distribution of all possible samples of two units from stratum 1 and one from stratum 2 along with their sample means.

| S. No. | Units in sample | Sample mean |
|---|---|---|
| 1 | ABC | 436.67 |
| 2 | ABE | 503.33 |
| 3 | ADC | 450.00 |
| 4 | ADE | 516.67 |
| 5 | AFC | 440.00 |
| 6 | AFE | 506.67 |
| 7 | BDC | 453.33 |
| 8 | BDE | 520.00 |
| 9 | BFC | 443.33 |
| 10 | BFE | 510.00 |
| 11 | DFC | 456.67 |
| 12 | DFE | 523.33 |

Let us compare samples presented in Tables 2.1 and 2.4 – one selected using SRS and the other using stratified random sampling. It can be checked that the mean of the 12 possible sample means in Table 2.4 is also equal to the population mean (480). It is also apparent that the 12 possible samples are scattered quite closely to the population mean. The sampling variance may be computed as:

$$\text{Sampling variance of mean} = \frac{(436.67-480)^2+\cdots+(523.33-480)^2}{12} = 1162.96$$

a drastic reduction in comparison to 67953.33 obtained earlier in SRS (Table 2.1).

Through stratification, therefore, it is possible to impose necessary restrictions so that the units in the sample, instead of belonging to few relatively homogeneous portions, are better spread in terms of capturing heterogeneity in a population. Stratification has the potentials to reduce the sampling variance of an estimate, compared to that in SRS design.

Let us elaborate it further. As mentioned, the sampling variance of mean in SRS (equation 2.9), depends mainly on $S^2$. Now,

$$(N-1)S^2 = \sum_{i=1}^{N}(y_i-\bar{Y})^2$$

In stratified sampling with L strata, the total variation can be written as

$$\begin{aligned}(N-1)S^2 &= \sum_{h=1,\,i=1}^{L,\,N_h}(y_{hi}-\bar{Y}_h+\bar{Y}_h-\bar{Y})^2 \\ &= \sum_{h=1,\,i=1}^{L,\,N_h}(y_{hi}-\bar{Y}_h)^2+\sum_{h=1,\,i=1}^{L,\,N_h}(\bar{Y}_h-\bar{Y})^2\end{aligned} \tag{2.22}$$

The product term, $2\sum_{h=1,i=1}^{L,N_h}(y_{hi}-\bar{Y}_h)(\bar{Y}_h-\bar{Y})$ is zero and $N_h$ is the number of units in $h$th stratum. If $S_h^2$ denotes the variance of elements in $h$th stratum, then

$$S_h^2 = \frac{1}{N_h - 1}\sum_{i=1}^{N_h}(y_{hi}-\bar{Y}_h)^2$$

The equation 2.22 can be written as

$$(N-1)S^2 = \sum_{h=1}^{L}(N_h - 1)S_h^2 + \sum_{h=1}^{L}N_h(\bar{Y}_h - \bar{Y})^2$$

Dividing both sides by $N$ and assuming that $\frac{1}{N}$ is negligible,

$$S^2 = \sum_{h=1}^{L}\frac{N_h}{N}S_h^2 + \sum_{h=1}^{L}\frac{N_h}{N}(\bar{Y}_h - \bar{Y})^2 \tag{2.23}$$

It will be discussed in Section 2.3.5.1 that only the first term on the right-hand side of equation 2.23 represents the sampling variance of mean for a stratified random sampling and hence is expected to be lower than the sampling variance of mean in SRS.

The gain in stratification would depend generally on how small is the value of

$$\sum_{h=1}^{L}\frac{N_h}{N}S_h^2$$

or alternatively how large is the value of

$$\sum_{h=1}^{L}\frac{N_h}{N}(\bar{Y}_h - \bar{Y})^2$$

In a stratified sampling, the endeavour is to subdivide the population into strata so that the variation among units within a stratum (denoted by $S_h^2$) is as small, that is, the units are as homogeneous as possible. To the extent it is possible to realize that a stratified random sampling can help in achieving considerable gain in reducing the sampling variance of an estimate.

## 2.3.2 Estimation of Parameters

In the stratified sampling, estimates are first prepared for each stratum. If stratified random sampling is used, that is, units within a stratum are selected by SRS, the estimation procedure outlined earlier in the case of SRS could be adopted for the stratum-level estimation. One would require to pool these stratum-level estimates suitably to get the overall estimate.

Let $N$ denote the size of a population that is subdivided into $L$ strata with $N_h$ units belonging to $h$th stratum ($h = 1, 2, \ldots, L$). $y_{hi}$ is the value of a characteristic for $i$th ($i = 1, 2, \ldots, N_h$) unit in $h$th stratum. Also, the total sample size $n$ is being distributed to the $L$ strata so that $n_h$ of the sample units fall in $h$th stratum.

$$n = n_1 + n_2 + \cdots + n_h + \cdots + n_L$$

In other words, $n_h$ units are selected independently and at random with a procedure similar to SRS from $N_h$ units in $h$th stratum.

#### 2.3.2.1 Estimation of mean

$\bar{Y}$, the population mean of the characteristic $y$ can be obtained as

$$\bar{Y} = \frac{\sum_{h=1}^{L} N_h \bar{Y}_h}{N} \tag{2.24}$$

where $\bar{Y}_h = \dfrac{\sum_{i=1}^{N_h} y_{hi}}{N_h}$

It is known that the sample mean of $h$th stratum $(\bar{y}_h)$ will be an unbiased estimate of the corresponding population value, $\bar{Y}_h$ (equation 2.3).

An unbiased estimate of the overall population mean $\bar{Y}$ can be obtained as

$$\bar{y}_{\text{st}} = \frac{N_1}{N}\bar{y}_1 + \frac{N_2}{N}\bar{y}_2 + \cdots + \frac{N_h}{N}\bar{y}_h + \cdots + \frac{N_L}{N}\bar{y}_L \tag{2.25}$$

where

$$\bar{y}_h = \frac{\sum_{1}^{n_h} y_{hi}}{n_h}$$

$\bar{y}_{\text{st}}$ is the sample estimate in the case of the stratified random sampling. The value of $\bar{y}_h$, being an unbiased estimate of the corresponding population mean in this case and hence can replace $\bar{Y}_h$ so that the two equations 2.25 and 2.24 would be equal.

Substituting the value of $\bar{y}_h$ in equation 2.25

$$\bar{y}_{\text{st}} = \frac{1}{N}\left[\sum_{h=1}^{L} \frac{N_h}{n_h}(y_{h1} + y_{h2} + \cdots + y_{hn_h})\right] \tag{2.26}$$

That is, each sample unit, $y_{hi}$ in $h$th stratum is multiplied by an inflation factor or weight $\dfrac{N_h}{n_h}$ and summed to get the numerator (i.e., an estimate of the population total) of equation 2.26.

It is known that

$$f_h = \frac{n_h}{N_h}$$

is the probability of selecting a unit in $h$th stratum. Consequently, the weight $(w'_h)$ for units in $h$th stratum is

$$w'_h = \frac{1}{f_h} = \frac{N_h}{n_h}$$

That is, the weight $w'_h$ is constant for all the $n_h$ units in $h$th stratum. It can be checked that if $w'_h$ is summed over all $n$ units in a sample and denoted as $w'$, then

$$\begin{aligned} w' &= \sum_{h,i} w'_h = \sum_h (w'_h + w'_h + \cdots + n_h \text{ times}) \\ &= \sum_h N_h = N \end{aligned}$$

Therefore, equation 2.26 can also be written as

$$\bar{y}_{st} = \frac{\sum_{h=1}^{L}\sum_{i=1}^{n_h} w'_h y_{hi}}{\sum_{h=1}^{L}\sum_{i=1}^{n_h} w'_h} \tag{2.27}$$

where $y_{hi}$ is the $i$th value of the variable $y$ in the $h$th stratum.

If values of $N_h$ are known, either of the two equations 2.26 or 2.27 can be used to estimate the mean of a stratified sample. If they are not known but an approximate measure of $f_h$ is used to determine the weights, equation 2.27 needs to be used for the estimation. It should be noted that, in such cases, equation 2.27 will give an approximate value of the estimate.

#### 2.3.2.2 Estimation of total

The estimate of population total can be obtained with the help of equation 2.26

$$\text{Estimated population total} = N \cdot \bar{y}_{st} \tag{2.28}$$

If $N$ is not known, equation 2.27 may be used

$$\text{Estimated population total} = w' \bar{y}_{st} \tag{2.29}$$

#### 2.3.2.3 Estimation of proportion

The overall proportion of an attribute in a population ($P$) is

$$P = \frac{\sum_{h=1}^{L} N_h P_h}{N}$$

where $P_h$ denotes the proportion in $h$th stratum.

If $p$ and $p_h$ are the corresponding sample values, the estimate of $P$ can be obtained as

$$p = \frac{\sum_{h=1}^{L} N_h p_h}{N} \tag{2.30}$$

Each individual ($y_{hi}$) can be given a weight ($w'_h$). $y_{hi}$ in this case is a dichotomous variable that takes a value of 1 if an individual possesses the attribute and 0 otherwise. Equation 2.30 can then be written as

$$p = \frac{\sum_h \sum_{i=1}^{n_h} w'_h y_{hi}}{\sum_h \sum_{i-1}^{n_h} w'_h} \tag{2.30a}$$

The estimate of total number of persons in the population with the attribute will be

$$\text{Estimated total with the attribute} = N \cdot p = \sum_{h=1}^{L} N_h p_h \tag{2.31}$$

### 2.3.3 Weighting and its Similarity with Standardization

It is pertinent to note, at this stage, the similarity between the use of weights in sample survey and the standardization technique often used for exercising control in impact evaluation by comparing two distributions (Section 7.4.1.2). This can perhaps be best understood in the context of stratification. Let us try to elaborate this with an example. Table 2.5 provides two separate distributions of prevalence of tobacco use and number of persons by age.

TABLE 2.5 Prevalence of tobacco use and number of persons by age in populations A and B.

| | Distributions in A | | Distributions in B | |
|---|---|---|---|---|
| Age group | Population (%) | Tobacco use rate (per 100) | Population (%) | Tobacco use rate (per 100) |
| 20–34 | 2070 (34.5) | 37.1 | 243 (40.5) | 38.5 |
| 35–59 | 2946 (49.1) | 45.6 | 217 (36.2) | 47.2 |
| 60 and above | 984 (16.4) | 25.2 | 140 (23.3) | 28.2 |
| Total | 6000 (100.0) | | 600 (100.0) | |

The purpose is to compare tobacco use rates for the two populations. For this, it is customary to obtain a single index like the prevalence of tobacco use in the two. They are

$$\text{Prevalence of use in A} = \frac{2070 \times 37.1 + 2946 \times 45.6 + 984 \times 25.2}{6000} = 39.3 \tag{2.32}$$

$$\text{Prevalence of use in B} = \frac{243 \times 38.5 + 217 \times 47.2 + 140 \times 28.2}{600} = 39.2 \tag{2.33}$$

The overall levels of prevalence are similar in the two. However, a look at the tobacco use rates by age reveals that they are quite different, and in fact, the rates are systematically higher in B compared to that in A. Why is then the overall prevalence similar in the two?

It is clear from the table that the tobacco use varies with age. Also, the age distributions in the two, which is required for obtaining the overall prevalence, are different. The factor age can, therefore, act as a confounder and disturb the comparison (this issue which is known as controlling the effect of a disturbing variable is discussed in part II of the book). The use of a standardization technique helps in holding the confounder, the age distribution in the present case as constant while making the comparison. Let us take age distribution of A as the standard and obtain the overall use rate in both A and B.

$$\text{Tobacco use rate in A} = 39.3 \text{ (as in equation 2.32)}$$

$$\begin{array}{l}\text{Tobacco use rate in B} \\ \text{(using age distribution of A)}\end{array} = \frac{2070x\ 38.5 + 2946x\ 47.2 + 984x\ 28.2}{6000} = 41.1 \tag{2.34}$$

Since the confounding factor, age is held constant in the two, it is now possible to compare their prevalence rates. The difference between the two can be ascribed as due to the other factor, the tobacco use (for a further discussion on the standardization technique, see Kitagawa (1964) and Roy and Unnikrishnan (1983)). This is the usual standardization technique.

What will happen if one of the two is a sample drawn from the other? Imagine that B is a sample drawn from A. Would it be alright to say that the sample estimate of prevalence (equation 2.33) which is an average of the sample values would provide a valid estimate of the population A? Does the age distribution difference in the two, which worked as a confounder in the comparison made earlier, matter in this case too? As discussed earlier, if the sample is drawn with the help of an EPSEM design, it will not matter. The age distribution difference will get reflected in the standard error of the estimate. Assuming that age is taken as a stratification variable, we can employ equation 2.27 to get the sample mean. In this case,

$$N_h = \text{Number of persons in } h\text{th age group in A, and}$$
$$n_h = \text{Number of persons in } h\text{th age group in B}$$

The variable $y_{hi}$ is a dichotomous variable indicating whether a person use's tobacco ($y_{hi} = 1$) or do not use ($y_{hi} = 0$). The prevalence of tobacco use in B ($p_h$) is, therefore,

$$p_h = \frac{\sum_{i=1}^{n_h} y_{hi}}{n_h} \text{ or } \sum_{i=1}^{n_h} y_{hi} = n_h p_h$$

where $p_h$ is the proportion (per cent) using tobacco in B.

To estimate the overall prevalence of tobacco use in B (the sample estimate), the weight for the three age groups ($h = 1, 2, 3$) would be $\left(\frac{2070}{243}\right)$, $\left(\frac{2946}{217}\right)$ and $\left(\frac{984}{140}\right)$ and according to equation 2.27,

the overall prevalence

$$= \frac{\left(\frac{2070}{243}\right) \times \sum_{i=1}^{243} y_{1i} + \left(\frac{2946}{217}\right) \times \sum_{i=1}^{217} y_{2i} + \left(\frac{984}{140}\right) \times \sum_{i=1}^{140} y_{3i}}{6000}$$

$$= \frac{\left(\frac{2070}{243}\right) \times 243 \times 38.5 + \left(\frac{2946}{217}\right) \times 217 \times 47.2 + \left(\frac{984}{140}\right) \times 140 \times 28.2}{6000} \quad (2.35)$$

$$= 41.1$$

It can be seen that the equations 2.34 and 2.35 are exactly same. Since age is a covariate, its distribution should be similar in a comparison, either between a population and a sample estimate or between two groups (generally referred to as a treatment and a control in evaluation designs, see Chapter 7). In sample survey, use of weights help in restoring the population distribution ($n_h$ is regarded as representing $N_h$). In standardization, the distribution of the covariate is kept as standard (constant). So, the concept of using weights in a survey design goes parallel to that of controlling the effect of confounders in an evaluation design (for example see Kish, 1987).

### 2.3.4 Sampling Variance of Estimates

As mentioned above, knowledge of sampling variance of estimates, in the case of SRS, comes in handy for the estimations in stratified sampling too.

#### 2.3.4.1 Sampling variance of mean

The estimated sampling variance of mean in $h$th stratum is, using equation 2.11

$$v(\bar{y}_h) = \frac{(N_h - n_h)}{N_h} \cdot \frac{s_h^2}{n_h}$$

where $s_h^2 = \dfrac{\sum_1^{n_h} (y_{hi} - \bar{y}_h)^2}{(n_h - 1)}$ is the sample estimate of the population mean square in $h$th stratum ($S_h^2$).

Since

$$\bar{y}_{st} = \sum_{h=1}^{L} \frac{N_h}{N} \cdot \bar{y}_h$$

$$v(\bar{y}_{st}) = \sum_{h=1}^{L} \left(\frac{N_h}{N}\right)^2 \cdot v(\bar{y}_h) = \frac{1}{N^2} \sum_{h=1}^{L} N_h (N_h - n_h) \frac{s_h^2}{n_h} \tag{2.36}$$

Substituting $N_h = w_h' n_h$, $v(\bar{y}_h)$ can be written as

$$v(\bar{y}_h) = \frac{(w_h' - 1)}{w_h'} \cdot \frac{s_h^2}{n_h}$$

Therefore,

$$v(\bar{y}_{st}) = \frac{1}{(\sum_{h,i} w_h')^2} \cdot \sum_{h=1}^{L} n_h w_h' (w_h' - 1) s_h^2 \tag{2.37}$$

#### 2.3.4.2 Sampling variance of total

Estimated population total $= N\bar{y}_{st}$

Therefore, variance of estimated population total $= N^2 v(\bar{y}_{st})$ (2.38)

#### 2.3.4.3 Sampling variance of proportion

The sampling variance of $p(v(p))$ is obtained as

$$v(p) = \sum_{h=1}^{L} \left(\frac{N_h}{N}\right)^2 v(p_h)$$
$$= \sum_{h=1}^{L} \left(\frac{N_h}{N}\right)^2 \cdot \frac{p_h q_h}{(n_h - 1)} \tag{2.39}$$

### 2.3.5 Allocation and Selection of Units

Allocation of units or the distribution of total sample size to different strata is an important dimension of stratified sampling. The gain in stratification depends on the formation of strata and the type of allocation of samples to different strata. That is, the gain depends on two distinct dimensions:

1. How the population should be subdivided into strata or to decide boundaries for each stratum, because once the boundaries for a stratum are fixed, population units that will belong to it will be known.
2. To decide how many units, out of the total samples to be allocated to different strata.

Gain in stratification is measured in terms of the reduction in sampling variance of an estimate in comparison to that of an SRS of same size. Considering the sampling variance of mean (equation 2.36), it can be noticed that its reduction depends on:

**(a)** the value of $S_h^2$ that is the element variance of a characteristic within the $h$th stratum, and

**(b)** sampling fraction in $h$th stratum or $\frac{n_h}{N_h}$

There are two different types of allocation. One is proportional allocation that considers the second aspect (point b above) while allocating samples to different strata, and the other is optimum allocation that takes into account both.

### 2.3.5.1 Proportional allocation

The sampling fraction is made constant in this allocation. That is, within the stratum $h$, the sampling fraction or the probability of selecting a unit, $f_h$ is constant. In other words,

$$\frac{n_h}{N_h} = \frac{n}{N} \tag{2.40}$$

It means,

$$n_h = \frac{n}{N} N_h \text{ or } n_h \ \alpha \ N_h \tag{2.41}$$

Further, equation 2.40 indicates that the probability of selecting a unit is same for all the units and it becomes an EPSEM design. It may be reiterated that like SRS, a stratified random sampling drawn with proportional allocation is also an EPSEM design, but unlike the former, all possible samples of size ${}^N c_n$ will not have equal chance of occurrence in the latter. As has been mentioned, giving equal chance to all possible samples may be undesirable, but giving equal chance for selection to each unit is desirable. Under the proportional allocation, by substituting

$$n_h = N_h \frac{n}{N}$$

in equation 2.36, the sampling variance of mean can be written as

$$v(\bar{y}_{st}) = \frac{(1-f)}{n} \sum_{h=1}^{L} \frac{N_h}{N} \cdot S_h^2 \tag{2.42}$$

Using equation 2.41, and denoting

$$\text{sampling variance of SRS} = v(\bar{y})_{\text{SRS}}$$

the equation 2.23 can be written as

$$v(\bar{y})_{SRS} = v(\bar{y}_{st}) + \frac{(1-f)}{n}\sum_{h=1}^{L}\frac{N_h}{N}(\bar{Y}_h - \bar{Y})^2 \tag{2.43}$$

Therefore, the relative gain in proportionate stratified sampling over SRS is obtained as

$$\text{relative gain in the stratification} = \frac{(v(\bar{y})_{SRS} - v(\bar{y}_{st})}{v(\bar{y})_{SRS}}$$

$$= \frac{\sum_{h=1}^{L}\frac{N_h}{N}(\bar{Y}_h - \bar{Y})^2}{S^2} \tag{2.44}$$

Few points emerge from the above discussions. The factor $\sum_{h=1}^{L}\frac{N_h}{N}\cdot S_h^2$ is the average value of the variance of elements in the L strata, and denoting it by $\bar{S}^2$, equation 2.42 can be written as

$$v(\bar{y}_{st}) = \frac{(1-f)}{n}\cdot\bar{S}^2 \tag{2.45}$$

The ratio of sampling variance of a design to that of the SRS of same size is known as the design effect. If the design effect of proportionate stratified random sampling is denoted by $\text{deff}_{\text{str}}$, then

$$\text{deff}_{\text{str}} = \frac{\frac{(1-f)}{n}\bar{S}^2}{\frac{(1-f)}{n}S^2} = \frac{\bar{S}^2}{S^2} \tag{2.46}$$

Higher the homogeneity between units within a stratum, lower will be the value of $v(\bar{y}_{st})$ (equation 2.45) relative to the sampling variance of SRS, which means higher the precision for the proportionate stratified sampling.

#### 2.3.5.2 Optimum allocation

The importance of $S_h^2$ can be gauged from the illustration shown in example 1.1 and discussed earlier. Putting units that are more or less homogeneous in one stratum can effectively reduce the values of $S_h^2$ and hence result in a gain in stratification.

Theoretically, it has been shown (Cochran, 1977) that the variance in the stratified sampling will be minimum if $n_h$ is selected such that

$$n_h = \frac{N_h S_h}{\sum_h N_h S_h} \text{ or } n_h \alpha N_h S_h \tag{2.47}$$

In this case, $n_h$ is decided on the basis of both the population size and standard deviation of the characteristic under study in *h*th stratum. This appears logical. Suppose a stratum has larger number of units consisting of daily wage labourer all getting similar wage. Since there is homogeneity among the units in terms of the characteristic, selection of few, even though there number is large, will suffice to capture the characteristic. It has been shown that the sampling variance in optimum allocation will be less than that in the proportional allocation (Cochran, 1977).

The above equation can also consider the cost aspect in a survey. It may be true that the cost of data collection in different strata vary considerably. If this is so, it is possible to take this into account, while allocating $n_h$, to make the design more efficient. The principle would be to have less samples from a stratum where the cost is high. For example, a population is subdivided into two strata. One is a hilly region where the cost of conducting the survey is twice that of the second which is a plain area. In addition to considering $N_h$ and $S_h$, that is, for given values of the two parameters, the sample size in the plains could be suitably increased compared to that in the hilly area. The equation 2.47 can be modified suitably (Hansen et al., 1953).

#### 2.3.5.3 Practical guidelines for allocation

Both proportional and optimum allocations serve important purposes. The former helps in having an EPSEM design. It also does not require much information for its implementation. The only information required is the size or estimated size of each stratum ($N_h$). The reduction in sampling variance of an estimate in optimum allocation can be substantial if either of the following holds:

- **(i)** if idea about the variability among units ($S_h^2$) is available,
- **(ii)** if it is possible to create strata in a manner so that $S_h^2$ can be regarded as similar (note that if $S_h^2$ are equal, equation 2.47 will be similar to that in the proportional allocation).

If the strata boundaries can be created in a manner so that $S_h^2$ are more or less equal, a proportional allocation can be as effective as the optimum allocation. The point remains that the values of $S_h^2$ will not be generally available and hence how to operationalize and use optimum allocation in an actual situation?

A large volume of research has been carried out to provide solutions in this regard (Cochran, 1961, Ekman, 1959 and Hansen et al., 1985). Few important guidelines emerge from such researches that facilitate the procedure of stratification, particularly for surveys with multiple objectives.

It has been found that the optimum value of $n_h$ is not very rigid. That is, a moderate departure from the optimum value can still be helpful in retaining the expected gain from the optimum allocation (Cochran, 1977 and Hansen et al., 1953).

Stevens (1952) suggested a method that advocates selection of equal sample size from each stratum. If $n$ is the sample size and there are $L$ strata, then according to this method

$$n_h = \frac{n}{L}$$

To reduce the sampling variance of an estimate, it is desirable to check variations in the values of $N_h$. Let, $N_{max}$ and $N_{min}$ are the maximum and minimum values of $N_h$. If ratio between the two is one, it means $N_h$ are exactly equal in all the strata. It is not necessary to make it exactly equal, but if the variation between the two is not much and restricted so that

$$\frac{N_{max}}{N_{min}} \leq 2 \tag{2.48}$$

the effectiveness of the method will be high.

The effectiveness will gradually decline as the ratio, $\frac{N_{max}}{N_{min}}$ increases beyond 2. Cochran (1977), while evaluating the precision of different methods of allocation, found that the precision of this method to be quite close to the optimum allocation. The implication of equation 2.48 is that stratum boundaries, to the extent possible, should be adjusted so that the size of each stratum is close to each other. It is possible to adjust the boundaries of strata and has been considered in the case of the National Family Health Survey (NFHS) in India (Roy and Pandey, 2008).

The method of drawing equal sample size from each stratum has few other advantages. There are practical advantages. It becomes easier to plan as well as to carry out the fieldwork, if sample sizes are equal. Administrative issues may also arise if unequal samples are drawn from different strata. If stratum sizes are unequal, selection of equal sample size from each of them will mean that the design will deviate from being EPSEM. If units within a stratum $h$ are selected with equal probability, that is,

$$f_h = \frac{n_h}{N_h}$$

then selected units in $h$th stratum will have the weight $w'_h \left(= \frac{1}{f_h}\right)$. If, in addition to having equal sample size in each strata, the stratum boundaries are also adjusted so that the values of $N_h$ are similar, apart from having a precision as high as optimum allocation, the design will also be close to a proportional allocation.

It may be mentioned that for estimating sampling variance (equation 2.37), it is necessary to have at least two units selected from each stratum. Keeping $n_h = 2$ has few advantages. The scope for stratification widens and it becomes possible to consider larger number of variables for stratification. This has been further discussed in Section 2.4.5.1.

### 2.3.6 Some Advantages of Stratification

Apart from improving precision of an estimate, stratification can serve few other purposes. It can facilitate estimation of parameters for some of the strata that are regarded as domains. Domain is a segment of a population for which independent estimation of the parameter of a study is required. A domain can be a stratum, if it satisfies the criteria of the stratification. Consider the objective of studying tobacco use in a population. It is found that a sample of 2000 individuals would be required to estimate the parameter with adequate precision. Further, 30 per cent of the population lives in urban and remaining 70 per cent in the rural areas. It is also known that the tobacco use behaviour is different in the two areas. For estimating the parameter for the overall population, one can consider residence (rural/urban) as a strata and allocate the sample preferably proportionally with 1400 (70 per cent) individuals in rural and 600 (30 per cent) in urban. However, if one requires estimation with similar level of precision separately for both rural and urban areas, in addition to the total, the sample size has to be increased and it will be around 4000 individuals with 2000 in each areas. In this case, the rural and urban areas will be referred to as domains. It is a nomenclature to remind that estimation is needed for a domain.

It needs to be noted that though sample size for estimating total population is 4000, its precision does not increase accordingly. Because the sample is not proportionally allocated to the two areas, the sampling variance of the estimate would be higher (because variations among weights will be higher to compensate for the disproportional allocation which will increase $\text{Meff}_w$, see equation 2.7) and inefficient to that extent than if it was done proportionally.

In stratified sampling, one can apply different designs for the estimation of parameters in different strata. There are pockets in a population that may be difficult to access for various reasons. Separate strata, in a stratified sampling, can contain such segments that are relatively inaccessible and can, for example, be sampled with a lower rate (which can be taken care of with weighting at the time of the estimation).

### 2.3.7 Post-stratification

The principle of stratification can be employed at the time of estimation even when a sample is selected without the stratification. This is known as post-stratification. Let us elaborate this in the context of estimation of tobacco use in a population. Age is generally a variable that is correlated with the use of tobacco. Hence, it is desirable that a sample selection procedure should consider age as a stratification variable for a bias-free estimation of the use. The age distribution of the population is available but for some reasons stratification was not done at the time of sampling. The plausible reason for not considering age could be that the information was not available at the level of the sampling units. An EPSEM design is used without the stratification. Let,

$N_i$ = number of persons from $i$th age group in the population
$N = \sum_i N_i$ = total size of population
$n_i$ = number of persons from $i$th age group selected in the sample
$n$ = total sample size
$P$ = proportion using tobacco in the population
$p_i$ = proportion of persons from $i$th age group using tobacco in the sample
$y_{ij}$ = a dichotomous variable indicating whether $j$th individual in $i$th age group is using tobacco $(y_{ij} = 1)$ or not $(y_{ij} = 0)$
$p$ = overall proportion using tobacco in the sample

Information on $n_i$ and $p_i$ can be easily obtained after interviewing the sample respondents. If the two distributions, $N_i$ and $n_i$ differ, instead of taking the simple average of $p_i$, one can employ the principle of stratification and obtain the estimated $p$ using equation (2.30) as

$$p = \frac{\sum_i N_i p_i}{N} \tag{2.49}$$

in the case of an EPSEM design. If it is not an EPSEM design, the estimate can be obtained as

$$p = \frac{\sum_{i,j} w'_i y_{ij}}{\sum_{i,j} w'_i} \tag{2.50}$$

where $w'_i = \frac{N_i}{n_i}$ is the weight for the $i$th age.

If reliable estimates of $N_i$ are available, the post-stratification can recover most of the advantages of stratification (Kish, 1995, p. 90–92).

## EXAMPLE 2.2

A rural population is divided into three strata based on their distance from the nearest town as i) within 3 km, ii) between 3 and 5 km, and iii) beyond 5 km. The number of households in the three strata is 500, 3000, and 6500, respectively. A total of 50 households, 10, 15, and 25, respectively, from each stratum were selected from the population using stratified random sampling and the values of monthly income (in rupees '000) are obtained as in Table 2.6.

TABLE 2.6 Stratum-wise monthly income.

| | Stratum 1 ($n_1 = 10$) | Stratum 2 ($n_2 = 15$) | Stratum 3 ($n_3 = 25$) |
|---|---|---|---|
| Monthly income (in rupees '000) | 10, 15, 50, 2, 8, 4, 18, 17, 7, 10 | 3, 12, 8, 9, 5, 2, 8, 7, 4, 11, 10, 5, 6, 3, 2 | 8, 7, 3, 4, 7, 10, 8, 9, 6, 6, 2, 5, 2 ,9 , 6 , 3, 3, 4, 6, 3, 2, 10, 3, 7, 8 |

Obtain the estimates of i) mean income and ii) proportion of persons having income less than Rs. 5000/- in the population. Also obtain the estimated standard error of the two estimates. The estimates of relevant variables for each stratum are presented in the Table 2.7.

TABLE 2.7 Stratum-wise different parameters and estimates

| | Stratum 1 | Stratum 2 | Stratum 3 |
|---|---|---|---|
| Population size ($N_i$) | 500 | 3000 | 6500 |
| Sample size ($n_i$) | 10 | 15 | 25 |
| Weights ($w'_i$) | $\frac{500}{10} = 50$ | $\frac{3000}{15} = 200$ | $\frac{6500}{25} = 260$ |
| Sample mean ($\bar{y}_i$) | $\frac{10+15+\cdots+7+10}{10} = \frac{141}{10} = 14.100$ | $\frac{3+12+\cdots+3+2}{15} = \frac{95}{15} = 6.333$ | $\frac{8+7+\cdots+7+8}{25} = \frac{141}{25} = 5.640$ |
| Mean squares ($s_i^2$) | $\frac{(10-14.1)^2+\cdots+(10+141.1)^2}{10-1} = 186.990$ | $\frac{(3-6.333)^2+\cdots+(2-6.333)^2}{15-1} = 10.667$ | $\frac{(8-5.64)^2+\cdots+(8-5.64)^2}{25-1} = 6.823$ |
| Proportion with income less than Rs. 5000 ($p_i$) | $= \frac{2}{10} = 0.200$ | $= \frac{5}{15} = 0.333$ | $= \frac{10}{25} = 0.400$ |

The pooled estimate of mean income of the population =

$$\frac{500\times 14.1+3000\times 6.333+6500\times 5.64}{10000}=6.271$$

Sampling variance of the mean income (equation 2.36) =

$$\frac{1}{10000^2}[500\times 49\times 186.99+3000\times 199\times 10.667+6500\times 259\times 6.823]=0.2243.$$

Estimated standard error of the mean = $\sqrt{0.2243}=0.4736$

The pooled estimate of the proportion with income less than Rs. 5000 ($p$) =

$$\frac{500\times 0.2+3000\times 0.333+6500\times 0.4}{10000}=0.37.$$

The estimated variance of $p$, $v(p)$ (according to equation 2.39)

$$\frac{1}{100000^2}[500^2(0.2\times 0.8/9)+3000^2(0.333\times 0.667/14)+6500^2(0.4\times 0.6/24)]=0.0057.$$

The standard error of $p=\sqrt{0.0057}=\ 0.075$.

## 2.4 SYSTEMATIC SAMPLING

### 2.4.1 Description

Instead of drawing individual random numbers for selection of each unit in an SRS, in systematic sampling, the selection is done, irrespective of the sample size $n$, with the help of a single random number. This is implemented by selecting only the first unit at random. The others get selected automatically with a fixed interval. This simplification in the selection used to be a distinct advantage before the era of computers. Nevertheless, the systematic manner in which a sample is selected generally turns out to be advantageous in this type of sampling. This technique is widely used in selection of samples, particularly at the last stage of selection, in a multistage design (discussed in Chapter 3).

### 2.4.2 Method of selection

The initial step in the selection is to obtain an interval of selection ($I$) based on sampling fraction ($f$)

$$I=\frac{1}{f}=\frac{N}{n} \tag{2.51}$$

Let $r$ denote a random number between 1 and $I$. Then $n$ samples are selected corresponding to:

$$r, r+I, r+2I, \ldots, r+(n-1)I$$

If $N$ units in a population are written serially in a list, then starting with $r$ every $I$th unit thereafter are automatically included in a sample.

For example, if a sample of size 4 is to be selected from a population of 20, then

$$I = \frac{20}{4} = 5$$

If $r$ is 2, the four samples selected are 2, 7, 12, 17.

### 2.4.2.1 Decimal interval method

The interval, $I$, can even be a fraction. If so, it can create a problem. For example, let $N$ is 26 and $n$ is 4.

$$I = \frac{26}{4} = 6.5 \sim 7$$

If $r = 1$, the samples will be 1, 8, 15, 22. But if $r = 7$, the samples will include 7, 14, 21, that is, 3 (because 21 + 7 = 28 does not exist). The sample size will, therefore, become a variable and can introduce bias. But if $I$ is large, greater than 5, the extent of bias can be ignored. That is, if the interval is large, the method of rounding it to nearest whole number and selecting the units as discussed is all right. Otherwise, one can use decimal interval method.

To illustrate, $I$, in the above example, can be taken as 6.5. The selection process can start with two digit numbers between 01 and 65 and ignoring the last digit for the selection. Let the two-digit random number be 65. Therefore, the four numbers to be used for the selection will be

$$r = 65 = 6.5 = 6 \text{ (ignoring the fraction, that is, rounding downward)}$$
$$r + I = 65 + 65 = 130 = 13$$
$$r + 2I = 65 + 130 = 195 = 19.5 = 19$$
$$r + 3I = 65 + 195 = 260 = 26.0 = 26$$

If $r$ is 01, the units selected will be

$$r = 01 = 0.1 = 0 \text{ (not considered)}$$
$$r + I = 01 + 65 = 66 = 6$$
$$r + 2I = 01 + 130 = 131 = 13$$
$$r + 3I = 01 + 195 = 196 = 19$$
$$r + 4I = 01 + 260 = 261 = 26$$

## 2.4.3 Advantages of Systematic Sampling

Apart from the ease of computation, systematic sampling can have few other advantages:

A list of sampling units may have some inbuilt ordering. For example, in a car registration, the units (cars) are registered as and when cars are purchased and hence provide a time series of the list. Time can be a relevant variable for studying a characteristic based on such register. If selected systematically, a sample can automatically take into account stratification by time. That is, it can capture the heterogeneity in the study variable

$y$ that is related to time. This is also known as **implicit stratification**. The households in an area are generally arranged serially in a geographical area. A systematic selection of households from such a list can provide some useful stratification, discussed later. The implicit stratification, obtained with the help of selecting samples systematically from an ordered list, does not impose any constraint on the total number of strata.

In some cases, the size of units in a population, $N$ may not be exactly known. It would be possible to draw a sample by fixing a sampling fraction and using systematic sampling. For example, one needs to select a sample of pages from a big file. Without counting the total pages, one can select a 10 per cent sample pages by systematic sampling. Since

$$I = \frac{1}{0.10} = 10$$

a random number between 1 and 10 (say, 6) can be selected. The sample will then consist of pages numbering 6, 16, 26, 36 and so on. Selection by a simple random sample of size $n$ will have a problem. Suppose the actual value of $N$ is 560 and it is approximated as 500. If an SRS is drawn, 60 pages (between 501 and 560) will not get a chance of being included in the sample. This can introduce substantial bias into an estimate. If it is drawn systematically with a random start, every unit will have a chance of being included in a sample.

For correct implementation of a sample design in the field, it is necessary to give clear and simple instructions to the field teams. Instructions should also be easy to supervise. A field team also has to make selections while they are in the field. A systematic selection often assists in ensuring randomness in the selection.

## 2.4.4 Disadvantages of Systematic Sampling

### 2.4.4.1 Monotonic trend

Some ordering in a list, as mentioned, can be beneficial for systematic selection. However, if the ordering is too strong and depicts a monotonic trend in $y$ values, systematic selection can be disadvantageous. To elaborate, let us consider an example of estimating average IQ of students in a class. There are 20 students in the class from which 4 are to be selected for the measurement. Suppose the list of students is such that there is a monotonic decline in the level of intelligence of each student. That is, $y_i$ (intelligence of the $i$th student) declines consistently with $i$. If selection is done systematically with an interval of 5, the five alternative samples are $(y_1, y_6, y_{11}, y_{16})$, $(y_2, y_7, y_{12}, y_{16})$ ... $(y_5, y_{10}, y_{15}, y_{20})$. The estimate of IQ will be the highest if the first sample is selected and lowest in fifth. Clearly, the sampling variance of the estimate will be high. Generally, such monotonic trends do not exist in a population. If it exists and is known, a sample that is placed in the middle will be preferred. In the present case, $(y_3, y_8, y_{13}, y_{18})$ is the one that is likely to be closest to the average IQ of the population. Since $y_i$ values are related, corrections can be done at the stage of estimation instead of at selection (Hansen et al., 1953).

### 2.4.4.2 Periodicity

There can be periodicity in the values of $y_i$ that, if present, can be more damaging. If decline is not monotonic but cyclic with a period of 5, as in the above example, that is, there is a declining trend in $y$ from $y_1$ to $y_5$ but $y_6$ is again similar to $y_1$ and again declines till $y_{10}$ and so on, the units selected in a sample will be homogeneous. Suppose in a housing complex, there are four buildings, each having four flats, one suitable for each economic class (very poor, poor, average and above average) numbered serially as 1 to 4, then 5 to 9, etc. A sample of size 4 drawn systematically will have all four belonging to a specific economic class and hence will fail to capture the heterogeneity. Such periodicity is also rare. However, remedies are available to reduce the effect of periodicity (Kish, 1995, p. 121–122).

One possibility is to break the ordering at the time of selection. For example, if in the first building the ordering is poorest to the richest (very poor as 1, poor as 2, so on), in the second building, the numbering can be reversed (above average as 5, and so on with very poor as 8), the third will again have like in the first, and the process of numbering will alternate similarly in the successive building. It is apparent that the complete homogeneity, all four sample units belonging to one economic category, will not arise in this case. A still better option, if periodicity occurs with such regularity, is to have the numbering within a building done randomly. That is, say, decide by lottery method whether a flat meant for very poor will get the starting number or any other.

Another alternative could be to have the selection not with one random number but with more than one number. For example, one can select $\frac{n}{2}$ samples with one random number and the remaining with another. For example, if $N = 400$ and $n = 40$. Since the interval $I$ is

$$I = \frac{400}{40} = 10$$

a random number between 1 and 10 can be selected. If the number is 3, one can select the first 20 units as (3, 13, 23, ..., 193). One can then select a fresh random start by selecting a number between 201 and 210. If this second number is 208, the next 20 units in the sample will be (208, 218, ..., 398).

### 2.4.5 Estimation of Parameters and Their Sampling Variances

The probability of selecting a unit is same for all the units and is equal to

$$\text{Probability of selecting a unit } = \frac{1}{I} = \frac{n}{N}$$

It is, therefore, an EPSEM design. The sample mean, like SRS is an unbiased estimate of the population mean. Similarly, the other parameters such as population totals and proportions can be estimated as done in the case of SRS.

It can be seen that, in the case of systematic sampling, the population units are subdivided into $n$ groups each of size $I$. One unit is then selected from each group (1 to $I$, $I + 1$ to $2I$, ..., $(n - 1)I$ to $nI$). That is, only one unit is selected, as if, from each stratum. At least two units are to be selected within a group (say, $h$) to estimate the value of $s_h^2$ and hence it is not possible to estimate the sampling variance of an estimate in the case of systematic sampling. However, there are ways to approximate sampling variance of an estimate (Kish, 1995, p. 117–119). Sampling variance of mean is only presented, the other parameters such as population total, proportions, etc., can be estimated similarly.

#### 2.4.5.1 Two consecutive units per stratum

Assuming two consecutive selected units to belong to one stratum of size 2$F$ (where $F = \frac{N}{n}$), a systematic sample of size $n$ can be regarded as selection of two units from $\frac{n}{2}$ strata. Following the principle of stratified sampling (equation 2.25)

$$\bar{y}_{\text{syst}} = \sum_{h=1}^{n/2} \frac{N_h}{N} \bar{y}_h \tag{2.52}$$

is an unbiased estimate of the population mean $\bar{Y}$, where $\bar{y}_h$ is the sample mean of $h$th stratum, having two observations, $y_{h1}$ and $y_{h2}$. That is,

$$\bar{y}_h = \frac{(y_{h1} + y_{h2})}{2} \tag{2.53}$$

As mentioned, each implicit stratum (addition of two consecutive intervals) is of size 2$F$ and hence

$$\frac{N_h}{N} = \frac{2F}{N} = \frac{2}{n} \tag{2.54}$$

With the help of equations 2.53 and 2.54, equation 2.52 can be written as

$$\bar{y}_{\text{syst}} = \sum_{h=1}^{\frac{n}{2}} \frac{(y_{h1} + y_{h2})}{n} \tag{2.55}$$

$$\begin{aligned} v(\bar{y}_{\text{syst}}) &= \sum_{h=1}^{\frac{n}{2}} \left(\frac{N_h}{N}\right)^2 v(\bar{y}_h) \\ &= \sum_{h=1}^{\frac{n}{2}} \left(\frac{N_h}{N}\right)^2 \frac{N_h - n_h}{N_h} \frac{s_h^2}{n_h} \end{aligned} \tag{2.56}$$

Now, since $n_h = 2$, and using equation 2.53

$$\begin{aligned} S_h^2 &= \frac{1}{(2-1)}\left[\left(y_{h1} - \frac{(y_{h1} + y_{h2})}{2}\right)^2 + \left(y_{h2} - \frac{(y_{h1} + y_{h2})}{2}\right)^2\right] \\ &= \frac{(y_{h1} - y_{h2})^2}{2} \end{aligned} \tag{2.57}$$

With the help of equations 2.54 and 2.57, from equation 2.56

$$\begin{aligned} v(\bar{y}_{\text{syst}}) &= \sum_{h=1}^{\frac{n}{2}} \left(\frac{2}{n}\right)^2 \frac{(2F-2)}{2F} \frac{(y_{h1} - y_{h2})^2}{4} \\ &= \frac{(1-f)}{n^2} \sum_{h=1}^{\frac{n}{2}} (y_{h1} - y_{h2})^2 \end{aligned} \tag{2.58}$$

## 2.5 PROBABILITY PROPORTIONAL TO SIZE SAMPLING

### 2.5.1 DESCRIPTION

When sampling units are of varying sizes, it is often advantageous to select them with differing probabilities related to their size. This is known as PPS sampling. Generally in PPS, sampling units refer to macro-level units like villages, towns, business establishments, schools, etc. PPS sampling can become appropriate while estimating a characteristic of individuals in two different situations:

1. The characteristic refers to macro-level units in a population. For example, estimating proportion of persons residing in villages (macro units) having certain facility like schools, public health centres, banks, etc., which are meant for everyone within the macro unit.
2. The characteristic to be estimated refers to individuals in a population without reference to any macro unit. This is the usual case of estimating socio-economic, health status, etc. of persons in a population.

It may be noted that macro units may assist in selection of individuals in both the cases. Also, in both the cases, the characteristic of interest refers to micro units, albeit with a difference. In case 1, all individuals within a macro unit will have similar characteristics. Suppose it is to study the availability of a bank facility in villages in a population or more specifically the proportion of population living in villages that have a bank facility. If a village has the facility, all its residents would be regarded as having the same. The cost of collection of information about the presence of a bank, in a selected village, does not depend on its size (number of its residents). On the other hand, if the objective is to study the proportion of persons having a bank account in a population, we have a situation as mentioned in case 2. Individuals within a village, unlike in case 1 can differ in terms of whether they have a bank account or not. Also, the information needs to be collected from each individual (sample of individuals) and hence, the size of a village will influence the cost.

Consider an area having $A$ villages with $V_i$ individuals in $i$th village. $V\left(=\sum_{i=1}^{A} V_i\right)$ denotes the total number of individuals in the area. Let us consider studying the availability of schools among the villages. Also assume that $K$ of the $A$ villages have schools.

**Case 1**: It is pertinent to note that there can be two different objectives in this case. The interest may be to estimate what proportion of villages ($A$) have schools ($P$). That is, to estimate $P$ where

$$P = \frac{K}{A}$$

Treating $A$ villages as the population units, one can select a required number of villages using any of the methods indicated earlier and estimate the sample proportion. Since the parameter to be estimated disregards the size of villages, PPS sampling is not needed.

Another aim can be to estimate the proportion of individuals staying in villages having a school ($P'$). This refers to

$$P' = \frac{\sum_{i=1}^{K} V_i}{V}$$

It needs to be emphasized that $P'$ can be quite different from $P$. If, for example, there are few very large size (in terms of number of population) villages having schools and many small size villages not having a school, $P$ can be small showing that very few villages are equipped with a school. On the other hand, $P'$ can be large indicating that a large proportion of the population, in the area, live in villages having a school.

For selecting a sample of '$a$' villages, any method discussed earlier can be employed. Let us consider that an EPSEM design is used to select '$a$' villages from $A$. The weight $(w_i')$ for $i$th village is

$$w_i' = \frac{A}{a}$$

and $\sum_{i=1}^{a} w_i' = A$. If $y_i$ represents whether a village has a school ($y_i = 1$) or not ($y_i = 0$), the estimated proportion of villages having a school ($p$) will be

$$p = \frac{\sum_{i=1}^{a} w_i' y_i}{\sum_{i=1}^{a} w_i'} = \frac{\sum_{i=1}^{a} y_i}{a} \tag{2.59}$$

Its variance can be estimated, using equation 2.17, as

$$V(p) = \frac{pq}{a-1} \tag{2.60}$$

This, in selection without replacement, assumes that the fpc can be ignored. In the case of selection with replacement, same units can get selected two or more times. The repeated selections are considered in the estimate.

Denoting $y_{ij}$ as $j$th ($j = 1, 2, \ldots, V_i$) person in $i$th ($i = 1, 2, \ldots, a$) selected village has a school ($y_{ij} = 1$) or not (($y_{ij} = 0$), the estimate of proportion of persons living in villages that have a school ($p'$) can be written as

$$p' = \frac{\sum_{i=1,j=1}^{a,V_i} w_i' y_{ij}}{\sum_{i=1,j=1}^{a,V_i} w_i'}$$

$$= \frac{\sum_{i=1,j=1}^{a,V_i} w_i'(y_{i1} + y_{i2} + \cdots + y_{iV_i})}{\sum_{i=1}^{a} (w_i' + w_i' + \cdots V_i \ times)}$$

Since,

$$y_{i1} = y_{i2} = y_{iV_i} = y_i$$

$$p' = \frac{\sum_{i=1}^{a} V_i y_i}{\sum_{i=1}^{a} V_i} \tag{2.61}$$

It should be noted that $w_i'$ is constant for all the $V_i$ persons in village $i$. $p'$ is a ratio estimate, its denominator, like the numerator, is a variable that depends on which villages are being selected. It will not be an unbiased estimate of $P'$ but the magnitude of bias would be small. Computation of variance of $p'$ is discussed in the next section.

Selection of villages by PPS is particularly helpful when primary sampling unit (PSU) size varies widely (Kish, 1995, p. 418–420). The method of selection in the case of PPS is discussed with an example subsequently. The probability of selecting $i$th village is

$$\text{probability of selecting } i\text{th village} = \frac{aV_i}{V},$$

and

$$\text{weight } (w_i'') = \frac{V}{aV_i}$$

Therefore,

$$p' = \frac{\sum_{i=1,j=1}^{a,V_i} w_i'' y_{ij}}{\sum_{i=1,j=1}^{a,V_i} w_i''} = \frac{\sum_{i=1}^{a} \frac{V}{aV_i} V_i y_i}{\sum_{i=1}^{a} \frac{V}{aV_i} V_i}$$
$$= \frac{\sum_{i=1}^{a} y_i}{a} \tag{2.62}$$

This is an unbiased estimate of the parameter $P'$. Its variance can be computed, using equation 2.17, as

$$V(p') = \frac{p'q'}{a-1} \tag{2.63}$$

where $q' = 1 - p'$. Example 2.3 provides an illustration of the procedure of selection and estimation in the case of a PPS sampling.

Before selection of macro-level units, whether with an EPSEM design or with PPS, generally stratification is employed, particularly by using village size as a criteria for stratification. If stratification is done, estimates are obtained within each strata as per the design used and then they are pooled together as done in the case of estimation in a stratified sampling.

**Case 2**: This case refers to the estimation of characteristic of individuals without any reference to a macro-level unit. The role of macro units can arise only in sample selection. In such cases, a cluster sampling can be adopted. Clusters that are a suitable macro units can be selected either randomly or preferably (if size of the macro units are available) by PPS. Estimation in the case of cluster sampling is discussed in Section 2.6. Another widely used approach is to select a sample in two or more stages, known as multistage design. This is discussed in the Chapter 3.

### EXAMPLE 2.3

Sizes of 8 villages in a population along with whether they have a school or not are given below. Draw a sample of three villages by PPS (with replacement) and estimate the proportion of population living in villages having a school and its standard error.

| S. No. | Village size (no. of persons) | Whether village has a school (yes = 1, no = 0) | Cumulative total no. persons | Range of random no. for selecting village $i$ |
|---|---|---|---|---|
| 1 | 50 | 0 | 50 | 1–50 |
| 2 | 200 | 1 | 250 | 51–250 |
| 3 | 140 | 1 | 390 | 251–390 |
| 4 | 75 | 0 | 465 | 391–465 |
| 5 | 225 | 1 | 690 | 466–690 |
| 6 | 165 | 1 | 855 | 691–855 |
| 7 | 40 | 0 | 895 | 856–895 |
| 8 | 80 | 0 | 975 | 896–975 |
| Total | 975 | | | |

For selection with replacement, three random numbers between 1 and 975, that is, three-digit random numbers, are to be generated. The cumulative total number of households up to village $i$ ($i = 1, 2, \ldots, 8$) is required for the selection (shown in column 4). The size of a village decides the number of random numbers that could be favourable for selecting it. For example, to select village 4, there are 75 numbers ranging from 391, which is one more than the cumulative total up to village 3, to 465, the cumulative total up to village 4. Let the three random numbers be 093, 953 and 467. Since 093, lies in between 51 and 250 (both inclusive), village 2 will be selected. The next random number 953 means village number 8 will be selected and 467 lie in between 466 and 690. Hence, village 5 will be selected. The sample consists of all the persons in villages 2, 5 and 8.

It can be checked that probability of selecting a village is proportional to its size. The probability of selecting village 2 is

$$\frac{3 \times 200}{975}$$

whereas the probability of selecting village 8 is

$$\frac{3 \times 80}{975}$$

The weights for the three villages are $\frac{975}{3 \times 200}$, $\frac{975}{3 \times 225}$ and $\frac{975}{3 \times 80}$. Using equation 2.62,

$$p' = \frac{\left(\frac{975}{3\times 200}\times 200\times y_1 + \frac{975}{3\times 225}\times 225\times y_2 + \frac{975}{3\times 80}\times 80\times y_3\right)}{\left[\left(\frac{975}{3\times 200}+\frac{975}{3\times 200}+\ldots 200\text{ times}\right)+\left(\frac{975}{3\times 225}+\frac{975}{3\times 225}+\ldots 225\text{ times}\right)+\left(\frac{975}{3\times 80}+\frac{975}{3\times 80}+\ldots 80\text{ times}\right)\right]}$$

$$= \frac{\sum_{i=1}^{a} y_i}{3}$$

$y_1$, $y_2$, and $y_3$ represents whether villages 2, 5 and 8 have the schools or not.

$$p' = \frac{2}{3} = 0.67$$

Its variance can be obtained using equation 2.63, which is equal to 0.11.

## 2.6 CLUSTER SAMPLING

### 2.6.1 Description

Clusters, generally some areal units, are the sampling units, in this case. Once a cluster is selected, all the units belonging to it are automatically included in a sample. Where areal units are considered as the clusters, for example, if villages are the clusters, all the residents in a selected village are included in the sample. If cluster size is large, it is usually found to be inefficient to include all the elements within a selected cluster. Chapter 3 outlines the alternative designs, known as multistage designs, where selection does not stop at one stage and is continued further. Multistage designs are extremely popular and have more or less become the rule. Since a group of units or a cluster is selected at the initial stage of a multistage design, they are also referred to as cluster sampling. In this chapter, we discuss the one-stage cluster sampling.

The word 'stratum' connotes a group of entities that are deemed to be similar to each other in terms of a characteristic under study. The strata are formed purposely to facilitate sample selection. A cluster, on the other hand, is a group of units that exist naturally, generally residing in a common administrative area. The characteristic under study of individuals in a cluster may or may not be similar. To further elaborate about units within strata or clusters, let us consider an example. The daily income of nine persons in a population is as shown in Table 2.8.

Table 2.8 Daily income of 9 persons.

| S. No. | A | B | C | D | E | F | G | H | I |
|---|---|---|---|---|---|---|---|---|---|
| Income | 50 | 600 | 9000 | 80 | 6000 | 100 | 800 | 5000 | 400 |

Suppose the sample size is three. In terms of the characteristic, income, there are three layers of people and it is clear that a good sample that represents the population should preferably consist of one person from each layer, such as (A, B, C) or (D, E, I), etc. A procedure of selection that generates alternative samples like (A, D, F) or (C, E, H), which does not depict the heterogeneity should be avoided. In stratified sampling, the effort would be to form three strata (A, D, F), (B, G, I) and (C, E, H) and select one from each to capture the heterogeneity.

In cluster sampling, if the arrangement of three clusters are (A, B, C), (D, E, G) and (F, H, I), selection of one cluster and including all three units in it in a sample will work well. The problem is clusters are usually not made artificially. In fact, the use of naturally existing clusters happens to be the main advantage of a cluster sampling albeit with some disadvantage. The difficulty is units within a naturally existing clusters are likely to be similar or related. In the above example, those with similar levels of income will be more likely to be found together in a cluster. This tendency to stay in the same area may be because of affinity or affordability. It is not only income, their living standard, attitude and preferences are also likely to be related. Because of this phenomenon, known as 'clustering', selection of clusters and inclusion of all its elements automatically into a sample tend to make a design inefficient, particularly if size of clusters happen to be large. The advantage of using naturally existing clusters happens to be that it allows, because of the availability of a frame, to draw a probability sample.

#### 2.6.1.1 Preparation of artificial clusters

Naturally existing clusters are widely used. In a country-level survey, it is generally impossible to employ a probability sampling to select individuals at one stage. Unless an updated population register is available which takes into account births, deaths, immigrations and emigrations, drawing of a probability sample of individuals directly will not be feasible. Cluster sampling considering some macro-level administrative areas like state, districts, blocks, villages, etc. as the sampling units can help in providing each and every individual resident a chance of inclusion in a sample.

It is, however, possible to prepare artificial clusters. For example, the national censuses conducted in India every 10 years, prepare census enumeration blocks (CEB) consisting of approximately 150–200 households in each city. The blocks are constructed within each ward (the lowest level administrative unit) of a city in a manner so that each and every resident can be located in a block. There are few points that need to be kept in mind while forming artificial clusters. One is the boundary of a cluster. Since, if selected, all the units within a cluster are to be included, any ambiguity in its boundary can create error. Also the error can take a systematic form. All individuals situated near the boundary are more likely to be misclassified and they may have some distinct characteristics, like they could be more recent migrants to the area, etc. Natural boundaries, like wards in an urban area or a village in rural is well demarcated and people can be expected to know the name of the place that they are staying. Another point for caution is that smaller the size of clusters, higher could be the extent of clustering. For example, in a city, one can consider each building (consisting of several residential units) as a cluster. However, the extent of clustering among the relatively small group of people staying in a building is expected to be higher, compared to that of an area consisting of more buildings taken as a cluster. This could be because of similarity in their socio-economic conditions, frequent interpersonal communications, etc.

The cost of survey will generally increase with the reduction in cluster size. More number of clusters, for a fixed sample size, are required to be selected. The cost of data collection within a cluster is likely to be much less compared to that of covering more number of clusters (discussed further in Chapter 3). There can also be an increase in the cost of obtaining the required frame for selection of clusters. If buildings in a city are taken as a cluster, the list of all buildings in the city will have to be prepared (if not known), whereas if blocks are used as a cluster, the frame is likely to be readily available.

### 2.6.2 Method of Selection

Clusters when listed can be selected either by SRS (without or with replacement) or by systematic sampling. In both the cases, it will be an EPSEM design, that is, the probability of selection of a cluster, irrespective of its size, will be same for each cluster. However, it is preferable to use PPS for the selection of clusters. The procedure of selection of clusters using PPS has been elaborated in Section 2.5.1, with an example.

As has been mentioned and will be emphasized subsequently, it is beneficial to use stratification for the selection of clusters. Considering the data in example 2.3, for selection of villages, one can form two strata with villages having less than 100 individuals and 100 or more. This will mean that variation in the size of villages within a stratum will reduce relative to that in the total population. Another alternative that can give similar effect as stratification by size is obtained if selection is made systematically by ordering villages according to their size. This has been shown, using data from example 2.3, in Table 2.9.

Table 2.9 Selection of three villages (in example 2.3) after arranging them according to their size.

| S. No. of village | Village ordered according to size | Cumulative size of villages |
|---|---|---|
| 1 | 40 | 40 |
| 2 | 50 | 90 |
| 3 | 75 | 165 |
| 4 | 80 | 245 |
| 5 | 140 | 385 |
| 6 | 165 | 550 |
| 7 | 200 | 750 |
| 8 | 225 | 975 |
| Total | 975 | |

If three villages are to be selected, the interval ($I$) for selection is

$$I = \frac{N}{n} = \frac{975}{3} = 325$$

The three villages can be selected using following three numbers: $R, R+I, R+2I$, where $R$ is a random number between 1 and $I$. If $R$ is 183, the three numbers will be 183, 508 and 833.

Three selected villages would then be village number 4 with a population 80, village number 6 with 165 persons and village 8 with 225 persons. If $k$ villages are to be selected, $I$ can be calculated as $I = \frac{N}{k}$ and selection can be done with the help of $k$ numbers as $R$, $R + I, R + 2I, \ldots, R + (k - 1)I$.

### 2.6.3 Estimation of Parameters and Sampling Variances

Estimation of sample mean and its variance has been discussed by considering clusters of equal and unequal sizes.

#### 2.6.3.1 Clusters of equal size

Let there be '$A$' clusters each of size $M$ and '$a$' of them are selected randomly, so that

$$\text{probability of selecting a cluster} = \frac{a}{A}$$

Let $y_{ij}$ where $i = 1, 2, \ldots, A$; $j = 1, 2, \ldots, M$ is the study variable. The sample mean for $i$th selected cluster, denoted as $\bar{y}_i$ is

$$\bar{y}_i = \frac{\sum_{j=1}^{M} y_{ij}}{M}$$

The overall estimated sample mean, denoted as $\bar{y}$ is

$$\bar{y} = \frac{\sum_{i=1}^{a} \bar{y}_i}{a} \tag{2.64}$$

Since all the $M$ units are included in the $i$th selected cluster, there are no within-cluster variations. The only variation, in the sample estimate, possible is due to variations in clusters that are included in a sample. This, like $S^2$ in SRS can be denoted by $S_A^2$, where

$$S_A^2 = \frac{\sum_{i=1}^{A} (\bar{y}_i - \bar{Y})^2}{(A-1)} \tag{2.65}$$

$\bar{Y}$ the population mean can be calculated as in 2.64 based on all $A$ clusters

$$\bar{Y} = \frac{\sum_{i=1}^{A} \bar{y}_i}{A}$$

The sampling variance of $\bar{y}$ in cluster sampling ($v(\bar{y})_{\text{cls}}$), like SRS (equation 2.11), can be obtained (e.g., Kish, 1995, p. 164–165) as

$$v(\bar{y})_{\text{cls}} = \left(1 - \frac{a}{A}\right)\frac{S_A^2}{a} \tag{2.66}$$

The value of $S_A^2$ is generally not available and the estimated sampling variance in cluster sampling can be obtained as

$$\text{Est. } v(\overline{y})_{\text{cls}} = \frac{(1-f_a)}{a} S_a^2 \tag{2.67}$$

where

$$f_a = \frac{a}{A}$$
$$S_a^2 = \frac{\sum_{i=1}^{a} (\overline{y}_i - \overline{y})^2}{(a-1)} \tag{2.68}$$

#### 2.6.3.2 Clusters of unequal size

More often than not, clusters are not of equal size, particularly when natural clusters are used. When cluster size is unequal, the total sample size is not fixed but a variable. Every cluster in a population has a chance either equal (if selected with SRS or systematic sampling) or unequal (if selected with PPS) of inclusion in a sample. Once the total number of clusters to be selected is decided, the sample size will depend on specific clusters that are selected.

For estimation of a parameter, mean of a variable, it is customary to denote sample mean $(\overline{y})$ as

$$\overline{y} = \frac{\sum_{i=1}^{n} y_i}{n} = \frac{y}{n}$$

However, since $n$ is a variable, $\overline{y}$ is a ratio. Its variance has to be calculated differently as discussed below.

There are situations where a parameter of interest is a ratio of two variables. Ratios are commonly used in epidemiological studies. For example, number of persons suffering from a disease in two different groups. The groups can be a treatment and a control in an evaluation study or any other socio-economic groups. It is better to represent both an estimate and its parameter as a ratio of two variables as

$$r = \frac{y}{x} \text{ and the parameter, } R = \frac{Y}{X}$$

Using the denominator as $x$, instead of $n$ (which generally denotes a constant) has another subtle advantage (Kish, 1995, p. 183–184). Consider, for example, estimation of proportion suffering from a disease. A sample of size $n$ shows $d$ to be suffering from the disease. As discussed earlier, the proportion ($p$) suffering is calculated as

$$p = \frac{d}{n} \text{ and } v(p) = \frac{pq}{n}$$

This is true if $n$ is fixed. In the case of cluster sampling, even if each cluster is selected with EPSEM, the estimation will have to be done differently, particularly the variance term should consider the variance of $n$. Denoting it as a ratio $r = \frac{d}{x}$ might act as a reminder towards this.

The ratio

$$r = \frac{y}{x} = \frac{\sum_i y_i}{\sum_i x_i} = \frac{\bar{y}}{\bar{x}} \tag{2.69}$$

is, though not an unbiased estimate of the parameter $R$, considered as a good estimator of $R$.

The sampling variance of $r$ is generally estimated using a Taylor series expansion. It is an approximation whose validity depends on certain conditions. The details of theoretical development of the formula are avoided here and the discussion on this can be found in Kish (1995). The estimated sampling variance of $r$ is given by

$$v(r) = \frac{1}{x^2}[v(y) + r^2 v(x) - 2r \text{ cov } (yx)] \tag{2.70}$$

$$= \frac{1}{x^2}[v(\bar{y}) + r^2 v(\bar{x}) - 2r \text{ cov } (\bar{y}\bar{x})] \tag{2.71}$$

As mentioned, $v(r)$ defined above is an approximation. If $c_x$ denotes the coefficient of variation of $x$, that is,

$$c_x = \frac{\sqrt{v(x)}}{\bar{x}} \tag{2.72}$$

to gauge the extent of the approximation, it can be mentioned that the actual sampling variance of $r$ is

$$(v(r))_{\text{actual}} = v(r)(1 + c_x + c_x^2 + c_x^3 + \ldots) \tag{2.73}$$

The second factor on the right-hand side of the equation 2.73, which is a Taylor series, is assumed to be equal to 1 with the assumption that the value of $c_x$ is sufficiently small. It has been shown that the estimates of variance based on equation 2.70 or 2.71 is tolerable if

$$c_x < 0.2$$

and $v$ $(r)$ provides a good estimate if

$$c_x < 0.1$$

It is, therefore, important to understand the significance of $c_x$ and how to control the variation in the sample size considered for each selected cluster. This will be discussed in the next chapter.

#### EXAMPLE 2.4

From a cluster of 1000 blocks consisting of 20 households each, 30 were selected randomly to estimate the proportion of households having at least one child of school going ages. The required data on number of households having at least one child are obtained as:

15, 5, 6, 3, 18, 11, 9, 7, 0, 16, 4, 8, 8, 9, 10,
12, 7, 9, 16, 5, 2, 4, 10, 15, 4, 9, 9, 12, 4, 7

Estimate the proportion of households having at least one child of school going ages in the population and its standard error.

Let,

$y_i$ = Number of households having a child of school going ages in the $i$th block

$\bar{y}$ = Average number of households having a child of school going ages

$$= \frac{\sum_{i=1}^{30} y_i}{30 \times 20} = \frac{254}{600} = 0.4233$$

Using equation 2.68, variations among the sample clusters can be obtained as

$$S_a^2 = \frac{\left(\frac{15}{20} - 0.4233\right)^2 + \left(\frac{5}{20} - 0.4233\right)^2 + \cdots + \left(\frac{4}{20} - 0.4233\right)^2 + \left(\frac{7}{20} - 0.4233\right)^2}{29} = 0.0506$$

Hence, using the equation 2.67, the estimated variance can be computed as

$$\text{Est. } v(\bar{y})_{\text{cls}} = \frac{\left(1 - \frac{30}{1000}\right)}{30} \cdot 0.0506 = 0.001637$$

Estimated standard error of the cluster sample mean = 0.0405.

## 2.7 KEY POINTS

- Apart from giving each possible sample an equal chance of being selected, an SRS with or without replacement also leads to an EPSEM design where each unit in a population gets equal chance of being included in a sample.
- Weight of a unit selected in a sample is inverse of its probability of selection. An EPSEM design attaches equal weight to each selected unit and is known as a self-weighting design. In an SRS of size $n$ from a population of $N$ units, each selected unit in the sample will have a weight equal to $\frac{N}{n}$.
- MSE of an estimate, which provides an idea about the likely deviations of a sample estimate from the population parameter, depends on following three factors:
  **(i)** bias in the estimate and its extent,
  **(ii)** heterogeneity among units in the population, and
  **(iii)** variation among weights of the sample units.
- Use of weights reduces bias in an estimate and hence is always recommended while estimating a parameter. However, the use of weights that vary from each other (i.e., departure from an EPSEM design) would increase the error in an estimate.
- The sampling variance of an estimate in an SRS design depends mainly on the degree of heterogeneity among units in a population. Larger the heterogeneity (measured by

the index $S^2$, known as the population mean square) higher is likely to be the sampling variance of an estimate.

- Stratified sampling is an effective tool to assist in the reduction of sampling variance of an estimate. In stratified random sampling, the population is first subdivided into relatively homogeneous groups (strata) and then samples are drawn independently from each strata.
- If $n_i$ units are drawn giving equal chance to all the $N_i$ units in $i$th stratum, they will have weights equal to $\frac{N_i}{n_i}$
- A proportional allocation of the total sample, $n$ to different groups $\frac{n_i}{N_i} = \frac{n}{N}$ will make the design self-weighting EPSEM. The gain in stratification increases in the case of proportional allocation.
- Selection of equal number of samples from each stratum has, apart from having operational advantages, potentials to reduce the MSE of an estimate. A particularly helpful approach is to adjust the stratum boundaries, if they are flexible, in a manner that would allow having approximately equal population size in each stratum. Drawing equal samples from each such stratum helps minimize the MSE of an estimate.
- Systematic sampling is another technique that provides equal chance to each sampling unit for selection in a sample. It has the potentials to provide the effect of stratification if there is any ordering of units in a frame.
- However, if there exists a monotonic trend in the arrangement of units in a frame, application of systematic sampling can be disadvantageous. This is also true when there is a periodic trend in the units in a frame.
- Systematic sampling is possible even when the total size of a population is unknown.
- If sampling units are clusters, PPS sampling is often a useful method to select sampling units. It, however, requires an idea about the size of each cluster.
- In cluster sampling, all units belonging to a selected cluster are included in a sample. Since units within a cluster generally have similarity in terms of the characteristic under study, known as clustering inclusion of all the units of a selected cluster into a sample makes the design less effective.

Chapter 3

# Multi-stage Designs

## 3.1 INTRODUCTION

Cluster sampling, as described Chapter 2, is a sampling technique in which all the units of a selected cluster are included in the sample. Because units in a cluster tend to be related to each other, the utility of a cluster sample reduces. Selection of all the units from a cluster, particularly if it is large in size curbs the ability to spread a sample and capture larger variation from the population under study. However, as we will see later, the effect of clustering tends to be inversely proportional to the cluster size. In other words, smaller the size of clusters, greater is the intra-cluster correlation. For example, consider clusters consisting of four or five consecutive houses in a village or residential flats in an urban dwelling unit. Because of the closeness between units in an area, it is expected that their social contacts and interpersonal communications will be higher among them and hence, apart from some socio-economic characteristics, many of their behaviours, attitudes, etc. are also likely to be similar giving rise to intra-cluster correlation among the individuals.

While in cluster sampling it is imperative to select all units within a cluster, it is often not viable to do so and as seen above, selecting all units from the cluster does not necessarily constitute an efficient sample. One alternative to taking all units from a selected cluster is to take subsamples from the selected clusters. That is, instead of giving all units of a selected cluster certainty of inclusion in a sample, one can select few units from each of them. Thus, in this case, selection of units is done in two stages – first selection of clusters from a list of clusters and then within each selected cluster selection of population units. In fact, this design, where the selection of samples is done in more than one stage, is widely used, particularly in large-scale population-based surveys. Generally, the administrative units existing in a country such as states, districts in a state, towns/villages in a district, etc. are used as sampling units at first stage of selection. These areal units have well-defined boundaries, each and every individual person or household can be uniquely identified as a resident of an area, and use of such units can also help in obtaining a sample that gives each resident a chance of being included in a sample. To draw a sample at a country level, one can have several options for the selection. For example, one can consider states as first-stage units and select few of them. If districts are the second-stage units, few districts can be selected from each of the selected states. One can progress like this till a population unit (households or a member of a household) is selected. It is possible to skip or add any stages of selection. That is, selection can be done at two or more stages. What facilitates the choice is the availability of a suitable frame for selection of sampling units at a given stage and cost as well as the error in an estimate that would be expected. A sampling unit that is selected at the first stage is known as the primary sampling unit (PSU). The PSUs are important in a multistage design, since its analysis is done at this level. Multistage design, discussed in the present chapter, refers to a probability sample selected at two or more stages. While a variety of such designs are feasible, a few specific variations of it are generally suitable and widely used and are mainly considered here for discussion.

## 3.2 TWO-STAGE DESIGN WITH EQUAL SIZE CLUSTERS

This section discusses the two-stage sampling design including the selection of units at each stage, the procedure of estimation of a parameter and sample estimates of the likely error in an estimate of a parameter. Although selection of units is possible in more than two stages, a two-stage sampling design forms the backbone of multistage designs. For example, if $y$ represents an outcome variable being studied, irrespective of the stages of the selection, the analysis essentially requires the cumulated values of $y$ for each selected PSUs.

Suppose in a multistage design, districts are the PSUs and 50 of which are selected. From each selected district, 9 villages are selected and in each selected village 20 households are included in a three-stage design to estimate average household size. Let $y_{ijk}$ denote the number of persons in $k$th household located in $j$th village of $i$th district. The analysis of sample mean and variance would require the values of $y_i$ for each selected PSUs, where

$$y_i = \sum_{k=1}^{20} y_{i1k} + \sum_{k=1}^{20} y_{i2k} + \cdots + \sum_{k=1}^{20} y_{i9k}$$

In other words, each $y_i$ will be a sum of 180 values of $y_{ijk}$. Because of this, PSUs are often referred to as 'ultimate clusters' or simply clusters. We will use 'PSU' and 'clusters' interchangeably, unless otherwise mentioned.

Examples of what may be considered as cluster vary widely and depend largely on the outcome of interest. For a household survey, for example, naturally existing areas are usually considered, such as the administrative areas like state, district or urban ward/village, as discussed earlier. Similarly, for survey of students, schools or classes within schools can be considered as natural clusters; for surveys regarding a disease, hospitals may be used as a cluster; for estimating total mangoes (or any fruit) in an orchard, mango trees can form the clusters; for estimating number of vehicles crossing a toll plaza, time may be used as cluster. While clusters of equal size greatly facilitates the statistical analysis, the natural clusters are rarely, if at all, of equal size. It is, however, possible to prepare clusters of equal size but that could be expensive. Many countries, particularly small in terms of population size, are subdivided into small blocks roughly of equal size. As population increases the number of blocks is changed to maintain the size of blocks. Even in a large country like India, at the time of a decadal census operation, areas are subdivided into enumeration blocks consisting of approximately 150–200 households in urban areas. Such enumeration blocks, known as CEB, are popularly used as urban frame in many surveys. This section discusses the procedure of selection at each stage in a two-stage design and analysis of data, mainly in terms of estimation of mean and its sampling variance.

### 3.2.1 Components of Overall Variation

In the context of a two-stage design, it is instructive to consider how the variance gets distributed in its two constituent parts. Suppose $N$ elements in a population are divided into '$A$' clusters each having '$B$' elements. Let $y_{ij}$ denote the value of a characteristic for $j$th element in $i$th cluster with $\bar{Y}$ as its overall mean and $\bar{Y}_i$ as the mean of $i$th cluster.

$$N = A \times B$$

Let, $S^2$ = Variations in '$y$' among elements in the population

$$S^2 = \frac{\sum_1^N (y_{ij} - \bar{Y})}{N-1}$$

Let us also define,

$$S_A^2 = \frac{\sum_{i=1}^A (\bar{y}_i - \bar{Y})^2}{A-1}$$

and,

$$S_B^2 = \frac{\sum_{i,j}^{A,B} (y_{ij} - \bar{Y}_i)^2}{A(B-1)}$$

where $S_A^2$ measures how each cluster mean differs from the overall mean.

In other words, it captures the variations between the clusters. It resembles $S^2$ but for the fact that $A$ cluster means are considered instead of $N$ elements. The second, $S_B^2$ measures how each element within a cluster differs from its mean. It provides the within-cluster variations. It can be checked that $B$ units within a cluster are used to calculate the within variation (element variance within a cluster) which is then averaged over the $A$ clusters. Hence, divisor is $A\ (B - 1)$.

Writing, $$\sum_1^N (y_{ij} - \bar{Y})^2 = \sum_1^N [(y_{ij} - \bar{Y}_i) + (\bar{Y}_i - \bar{Y})]^2$$

It can be easily shown that in a two-stage design

$$\left(\frac{(N-1)}{N}\right) S^2 = \left(\frac{(A-1)}{A}\right) S_A^2 + \left(\frac{(B-1)}{B}\right) S_B^2 \tag{3.1}$$

The implication of this relationship needs to be noted. For a given population, its value of $S^2$ is fixed. Hence, larger the variation between clusters ($S_A^2$), smaller will be the within-cluster variance ($S_B^2$). This means, if the formation of clusters is such that elements within a cluster tend to be alike in terms of $y$, then ($S_B^2$) will be small and given $S^2$, the between-cluster variation will naturally be large to that extent. This is expected in natural clusters. The ratio of $S_A^2$ to $S^2$ or $S_B^2$ to $S^2$ can indicate the extent of the clustering in a design. In fact, the intra-cluster correlation ($\rho$) can be computed as (Kish, 1995, p. 170–172)

$$\rho = 1 - \frac{S_B{}^2}{S^2} \times \frac{N}{N-1} \tag{3.2}$$

Idea about bifurcation of the total variation into its two constituent parts is further elaborated through an example below.

### Example 3.1

The monthly income (in Rs. '000) of 16 households, in a hypothetical population, is as follows:

| S. No. | 1 | 2 | 3 | 4 | 5 | 6 | 7 | 8 | 9 | 10 | 11 | 12 | 13 | 14 | 15 | 16 |
|---|---|---|---|---|---|---|---|---|---|---|---|---|---|---|---|---|
| **Income** | 5 | 7 | 8 | 3 | 20 | 25 | 21 | 30 | 55 | 60 | 48 | 46 | 96 | 80 | 76 | 92 |

Consider the following two different living arrangements of the 16 households in four clusters:

**(a)** Households with similar income living in the same cluster

| Cluster | Monthly income of households ($y$) | | | | Mean income ($\bar{y}_1$) $\bar{y}_i$ |
|---|---|---|---|---|---|
| 1 | 5 | 7 | 8 | 3 | 5.75 |
| 2 | 20 | 25 | 21 | 30 | 24.00 |
| 3 | 55 | 60 | 48 | 46 | 52.25 |
| 4 | 96 | 80 | 76 | 92 | 86.00 |

The overall mean monthly income of households in the population is

$$\bar{y} = \frac{(5+7+8+\cdots+76+92)}{16} = 42$$

Total element variance of income $S^2$:

$$S^2 = \frac{(5-42)^2+(7-42)^2+\cdots+(76-42)^2+(92-42)^2}{15} = 1012.67$$

$$S_A^2 = \frac{(5.75-42)^2+(24-42)^2+(52.25-42)^2+(86-42)^2}{3} = 1226.38$$

$$S_B^2 = \frac{1}{4}\left[\frac{(5-5.75)^2+(7-5.75)^2+\cdots+(3-5.75)^2+(20-24)^2+\cdots+(92-86)^2}{3}\right] = 39.46$$

It can be checked that equation 3.1 is satisfied and equation 3.2 provides a very high value of $\rho$ equal to 0.96, showing that population within each cluster is homogeneous in terms of income.

**(b)** Households with different income levels live in each cluster

| Area/cluster | Monthly income of households ($y$) | | | | Mean income ($\bar{y}_1$) |
|---|---|---|---|---|---|
| 1 | 5 | 20 | 55 | 76 | 39.00 |
| 2 | 7 | 25 | 60 | 80 | 43.00 |
| 3 | 8 | 21 | 48 | 92 | 42.25 |
| 4 | 3 | 30 | 46 | 96 | 43.75 |

$$S_A^2 = \frac{(39.0-42)^2+(43-42)^2+(42.25-42)^2+(43.75-42)^2}{3} = 4.38$$

$$S_B^2 = \frac{1}{4}\left[\frac{(5-39.0)^2+(20-39.0)^2+\cdots+(76-39.0)^2+(7-43)^2+\cdots+(96-43.75)^2}{3}\right] = 1261.46$$

One can again check the validity of equation 3.1 and calculate the value of $\rho$ from equation 3.2. The value of $\rho$ is found to be negative (–0.33), implying complete absence of clustering of population in terms of income.

From the above example, we note that in the presence of clustering, the value of $S_A^2$, the inter-cluster variation, is high and it decreases with decreasing amount of clustering (compare two situations in the above example). In other words, larger the degree of clustering, greater will be the contribution of $S_A^2$ to the total variation.

We illustrate below the computation using STATA software

```
/*
Example 3.1: The monthly income (in Rs. '000) of 16 households, in a hypothetical
population, is as follows:

 HH Number   1   2   3   4   5   6   7   8   9   10  11  12  13  14  15  16
 Income      5   7   8   3   20  25  21  30  55  60  48  46  96  80  76  92

Consider the following two different living arrangements of the 16 households
in four clusters:

a) Households with similar income living in the same cluster
```

```
    Cluster          Monthly income of households (y)
    1                 5             7              8              3
    2                20            25             21             30
    3                55            60             48             46
    4                96            80             76             92

b) Households with different income levels live in each cluster
    Cluster          Monthly income of households (y)
    1                5             20             55             76
    2                7             25             60             80
    3                8             21             48             92
    4                3             30             46             96

Problem: To show how intra-class correlation differs in two situations
Data first entered using STATA data editor (see below)

 id            income         clusid_1       clusid_2
 1               5              1              1
 2               7              1              2
 3               8              1              3
 4               3              1              4
 5              20              2              1
 6              25              2              2
 7              21              2              3
 8              30              2              4
 9              55              3              1
 10             60              3              2
 11             48              3              3
 12             46              3              4
 13             96              4              4
 14             80              4              2
 15             76              4              1
 16             92              4              3
 */

***calculate S^2 (total variance), Sa^2 (within cluster variance) and Sb^2
(between cluster variance for two different clustering

**for 1st form of clustering

loneway income clusid_1

[OUTPUT]

One-way Analysis of Variance for income: Income

Number of obs =  16
R-squared =      0.9688
```

```
Source                SS          df        MS                F          Prob > F

Between clusid_1      14716.5      3        4905.5            124.32     0.0000
Within clusid_1       473.5       12        39.458333

Total                 15190       15        1012.6667

Intraclass    Asy.
correlation   S.E.          [95% Conf. Interval]
-----------------------------------------------------
0.96858       0.02800       0.91370        1.02347

Estimated SD of clusid_1 effect           34.87851
Estimated SD within clusid_1              6.281587
Est. reliability of a clusid_1 mean       0.99196
(evaluated at n=4.00)

Note that STATA gives SS or MS instead of variance. Intra-class correlation is
estimated as 0.968

*** for 2nd form of clustering

loneway income clusid_2

[OUTPUT]
One-way Analysis of Variance for income: Income
                                                  Number of obs =          16
                                                      R-squared =      0.0035

Source                    SS        df       MS             F        Prob > F
------------------------------------------------------------------------------
Between clusid_2          52.5       3       17.5           0.01     0.9976
Within clusid_2        15137.5      12     1261.4583
------------------------------------------------------------------------------
Total                    15190      15     1012.6667

Intraclass       Asy.
correlation      S.E.       [95% Conf. Interval]
----------------------------------------------------
0.00000*     0.22822      0.00000      0.44730

Estimated SD of clusid_2 effect              .
Estimated SD within clusid_2              35.51701
Est. reliability of a clusid_2 mean        0.00000*
(evaluated at n=4.00)
(*) Truncated at zero.

Note: Since intraclass correlation is negative, its truncated at zero
```

Using the analysis of variance analogy, it has been shown that the variance of the sample mean $(v(\overline{y}))$ can be approximated as

$$v(\overline{y}) = \frac{S_A^2}{a} + \frac{S^2}{ab} \tag{3.2a}$$

where '$a$' denotes the number PSUs selected and '$b$' is the number of units selected in a selected PSU (Snedecor and Cochran, 1989). As will be discussed later also, the increase in sampling variance of mean in the case of multistage design depends mainly on $S_A^2$, and as evident from equation 3.2a, for a given sample size $n$ (= $a \times b$), it is beneficial to consider a larger number of PSUs (a) than increasing the number of units selected in a PSU (b).

### 3.2.2 Two-stage design for selection of units with equal probability

There can be a variety of two-stage designs. It is, however, desirable to have an EPSEM design (also known as self-weighting design), that is, a design that selects units from a population giving equal probability of selection to all its units. In this section, we discuss the procedure of selection of units as well as estimation of parameters in a two-stage design in which units are selected with equal probability.

#### 3.2.2.1 Procedure of selection in a two-stage EPSEM design

Let,

$a$ = number of clusters (PSUs) to be selected,
$b$ = number of elements to be selected from each selected cluster and
$n = a \times b$, the total number of elements or the sample size.
The sampling fraction $f$ is given as

$$f = \frac{n}{N}$$

The value of $f$ is the overall probability of selecting a unit from the population. Since the selection is made at two stages, $f$ is a product of $f_1$ and $f_2$, where,

$f_1$ = probability of selecting a PSU and
$f_2$ = probability of selecting a unit within a selected PSU.
In other words,

$$f = f_1 \times f_2 \tag{3.3}$$

To have an EPSEM or self-weighting design, units, at both the stages, can be selected with equal probability. We thus write,

$$f_1 = \frac{a}{A} \text{ where A is the total number of PSUs in the population}$$

$$\text{and } f_2 = \frac{b}{B}, \text{ where B is the total number of units in each selected PSU}$$

$$\text{Therefore, } f = \frac{a}{A} \times \frac{b}{B} = \frac{ab}{AB} = \frac{n}{N}$$

This means that at the first stage, '$a$' PSUs are selected from a total of A PSUs by giving all of them equal chance. Similarly, in the second stage, $b$ units are selected from each of '$a$' selected PSUs by giving equal chance to all B units in it.

In such a design, since $n$ and $N$ are available, one needs to decide either $f_1$ or $f_2$. The other can be obtained by using equation 3.3. For example, if the sampling fraction ($f$) is 5 per cent and one decides to select 10 per cent of first-stage units ($f_1$), then $f_2$ will be

$$f_2 = \frac{f}{f_1} = \frac{0.05}{0.10} = 0.50$$

The procedure of selection, in this case, would be

$a = 0.10$ A, that is, 10 per cent of the first-stage units in the population, and
$b = 0.50$ B, that is, 50 per cent of the second-stage units in the population.

It can be seen that by changing the values of $f_1$ or $f_2$, one can have several alternative designs. For example, if $N = 20{,}000$ consisting of 400 clusters (A) each having 50 individuals (B), a sample of 200 individuals (i.e., $f = 1$ per cent) can be obtained by selecting, say, 100 clusters ($f_1 = 25$ per cent) and then selecting 2 individuals ($f_2 = 4$ per cent) from each selected clusters. Alternatively, one can select 40 clusters ($f_1 = 10$ per cent) and 5 individuals ($f_2 = 10$ per cent) from each cluster. Alternative designs depend on the choice of a balance between number of clusters (PSUs) and sample to draw from each cluster. As indicated in Section 3.2.1, it is advisable to consider having as many PSUs as possible. We will discuss this further in subsequent sections.

Apart from deciding on the number of sampling units to be selected, one also needs to choose a selection procedure. One can choose between SRS with or without replacement and a systematic sampling to select PSUs in first stage and elements in a PSU at second stage. For selecting units at second stage, one can also use cluster sampling. In this case, the second-stage units can be suitably grouped into small clusters and few clusters, depending on the number of units to be selected, can be selected with EPSEM. A type of cluster sampling for selecting households at second stage, for example, was used in India Demographic and Health Survey (NFHS-3). If $b$ were the number of households to be selected at the second stage, $\frac{b}{5}$ households were first selected systematically and interviewed. In addition to interviewing a selected household, four more households that were adjacent to the selected household were also included in the sample and interviewed. That is, each selection provided a cluster of five contiguous households. Although this introduced some clustering effect (because all the $b$ households were not selected independently), this design was necessary to provide support to HIV testing of household members. Survey teams are required to carry necessary equipment to the selected households and using cluster sampling at the second stage helped reducing burden on the survey teams. In addition to logistical support, selection of contiguous households within a PSU helped reduce stigmatization of households in which HIV tests were to be done and thereby minimizing refusal and social desirability bias. Further discussion on selection of units at different stages can be found in Section 3.5.

#### 3.2.2.2 Estimation of mean and variance

If $y_{ij}$ ($i = 1, 2, \ldots, a$; $j = 1, 2, \ldots, b$) denotes a sample characteristic for $j$th selected unit in $i$th selected PSU, the sample mean ($\bar{y}$) can be obtained as:

$$\bar{y} = \sum_{ij} \frac{y_{ij}}{n} \tag{3.4}$$

If $y_i$ is the total of $b$ units selected in $i$th PSU,

$$\bar{y} = \sum_{i} \frac{y_i}{n}$$

Since sample size is fixed and selected with equal probability, $\bar{y}$ is an unbiased estimate of the population mean $\bar{Y}$,

$$\bar{Y} = \sum_{ij} \frac{y_{ij}}{N}$$

The summation is taken over all the units in A PSUs in the population. Assuming that the sampling units were selected at both stages without replacement, the estimate of sampling variance of $\bar{y}$ is given as

$$V(\bar{y}) = \left(1 - \frac{a}{A}\right)\frac{S_A^2}{a} + \left(1 - \frac{b}{B}\right)\frac{S_B^2}{ab} \tag{3.5}$$

This resembles the sampling variance of SRS of elements as mentioned in Chapter 2, except for the fact that, in the present case we have a component of variance for each of the first- and second-stage sampling. Since $S_A^2$ and $S_B^2$ are the population parameters and hence generally not available, for estimating sampling variance of $\bar{y}$, one needs to have their estimates from a sample.

It can be shown that $s_b^2$ is an unbiased estimate of $S_B^2$ and, therefore,

$$s_b^2 = \frac{\sum_{ij}(y_{ij} - \bar{y}_i)^2}{a(b-1)} \tag{3.6}$$

can be estimated from the sample values and used instead of $S_B^2$. However, a sample value of $s_a^2$ given as

$$s_a^2 = \frac{1}{(a-1)}\sum_{i=1}^{a}(\bar{y}_i - \bar{y})^2 \tag{3.7}$$

is not an unbiased estimate of $s_A^2$. It has been shown that $\left[s_a^2 - \left(1 - \frac{b}{B}\right)\frac{s_b^2}{b}\right]$ is an unbiased estimate of $s_A^2$ (Kish, 1995, p. 166–168). Hence, with the help of equation 3.5, it is possible to write the estimated sampling variance of $\bar{y}$ as

$$V(\bar{y}) = \left(1 - \frac{a}{A}\right)\frac{1}{a}\left[s_a^2 - \left(1 - \frac{b}{B}\right)\frac{s_b^2}{b}\right] + \left(1 - \frac{b}{B}\right)\frac{s_b^2}{ab} \tag{3.8}$$

The equation can be further simplified as

$$V(\bar{y}) = \left(1 - \frac{a}{A}\right)\frac{s_a^2}{a} - \frac{1}{A}\left[S_a^2 - \left(1 - \frac{b}{B}\right)\frac{s_b^2}{b}\right] \tag{3.9}$$

The above equation can be approximated (assuming that the contribution of the second component of equation 3.9 is negligible since the denominator A in the second term would be large) to

$$V(\bar{y}) = \left(1 - \frac{a}{A}\right)\frac{s_a^2}{a}$$

This can also be written (with the help of equation 3.7) as

$$V(\bar{y}) = \left(1 - \frac{a}{A}\right)\left(\frac{1}{a\,(a-1)}\right)\sum_{i=1}^{a}(\bar{y}_i - \bar{y})^2 \tag{3.10}$$

writing $\bar{y}_i$ (the mean of y values in $i$th PSU) as

$$\bar{y}_i = \frac{\sum_{j=1}^{b} y_{ij}}{b} = \frac{y_i}{b}$$

and,

$$\bar{y}(\text{the mean of } y \text{ values in all '}a\text{' PSUs}) = \frac{\sum_{i,j}^{a,b} y_{ij}}{ab}$$

equation 3.10 can also be written as

$$V(\bar{y}) = \left(1 - \frac{a}{A}\right)\left(\frac{1}{a(a-1)}\right)\frac{1}{b^2}\left(\sum_{i=1}^{a} y_i^2 - \frac{y^2}{a}\right) \tag{3.10a}$$

The actual estimated sampling variance computed on the basis of equation 3.9 will be lower compared to that if 3.10 is used. The equation 3.10 or 3.10a, normally used because of its simplicity as well as other advantage, will tend to overestimate the sampling variance, but the extent of the overestimation will only be marginal, particularly if number of clusters (A) in a population is large.

The advantage of equation 3.10 is that it estimates sampling variance on the basis of PSU (means and totals of each selected PSUs) alone and it is not necessary to know individual values of $y_{ij}$ for $j = 1, 2, \ldots, b$ in $i$th PSU. It also avoids examination of intra-cluster variations in $y(s_b^2)$. To obtain precision of the estimate, it is sufficient, though to get idea about the contributions of first- and second-stage units separately, one will need estimate of $s_b^2$. The computation of sampling variance from PSU means and totals, using equation 3.10 also remains valid in the cases where samples are selected using three or more stages of sampling.

#### 3.2.2.3 Clustering, design effect and choice of number of PSUs/cluster

It is informative to compare the variance in the case of a cluster design with that if units were selected directly (in one stage) using SRS. A ratio of the two variances is known as the 'design effect' (deff), that is,

$$\text{deff} = \frac{\text{sampling variance in cluster design of, say, } n \text{ units}}{\text{variance in the case of simple random sample of size } n} \tag{3.11}$$

This ratio can also be obtained as (Kish, 1995, p. 161–162)

$$\text{deff} = 1 + (b - 1)\rho \tag{3.12}$$

where $\rho$ denotes the intra-cluster correlation and $b$ is the cluster size (by cluster size we mean the number of units selected in a cluster or size of an ultimate cluster). The value of deff provides a comparison of relative precision of a cluster design to that of a simple random sample of similar size. It gives the extent of increase, if any, in the sampling variance because units are not selected in one stage using SRS but are drawn in two or more stages. The extent of increase depends on i) extent of intra-cluster correlation and ii) number of units (b) that are selected from each selected cluster.

From the equation 3.12 it is clear that for a given b, higher the intra-cluster correlation, greater will be the value of deff, which means the sampling variance of an estimate obtained with the help of a clustered design of size $n$ will be higher than a sample of size $n$ drawn by SRS. Naturally, higher the values of $\rho$, more is the similarity between units within a cluster (also indicated in Section 3.2.1). The presence of $\rho$ influences the ability to draw a sample of $n$ independently and identically distributed units, a primary assumption in most statistical tests (Kish and Frankel, 1974). This is because, although '$a$' PSUs are selected independently from a population, $b$ units that are selected in a PSU cannot be regarded as independent samples from the population. They may also be selected independently but with a condition that they should all belong to the same PSU. And the fact is units within a PSU are not likely to be distributed randomly as in the population. It has been shown that (Kish, 1995, p. 259),

$$n_{\text{eff}} = \frac{n}{\text{deff}} \tag{3.13}$$

where $n_{\text{eff}}$ is the effective sample size or the number of elements that can be considered to have been selected under the assumption that they are independently and identically distributed. As deff increases, for a given, $a$ and $b$, the effective sample size decreases. For example, if sample size ($n$) is 900 and a clustered design is used with a value of deff of 1.5, the effective sample size will be 600. It may be noted from equation 3.12 that if $\rho = 1$, the value of deff will be equal to $b$. Therefore, from equation 3.13,

$$n_{\text{eff}} = \frac{n}{b} = a$$

This shows that if there is total correspondence or there is no variation among units in a PSU, the effective sample size will be equal to the number of PSUs selected.

The value of design effect is utilized while deciding the size of a sample in a clustered design. Consideration of deff or $n_{\text{eff}}$ is also of significance in many evaluation designs based on units drawn from a population using a clustered design. If cluster size ($b$) varies, as will be discussed later, an average value of $b$ computed as $\left(\frac{n}{a}\right)$ can be used for it.

The value of $\rho$ and hence, for a given $b$, the value of deff is portable and can be used by surveys that proposes to use more or less similar designs. Because of this, it is generally advised to include value of deff as well as $\rho$ in a survey report. All Demographic and Health Survey (DHS) surveys provide information on design effects. The value of deff can be used to interpret the extent of clustering (equation 3.11) in a parameter or to understand the effective sample size (equation 3.13). The overall design effect (deff), generally presented in a survey, however, could be different. The value of deff indicated here refers to the increase in sampling variance of a parameter that happens in a design exclusively due to the effect of clustering. The sampling variance of a variable computed in a design, on the other hand, includes the changes in it that can occur due to a deviation from an EPSEM selection, adjustment due to the non-response, use of stratification with or without proportional allocation, etc. (Kalton et al., 2005). In many household-based surveys, it has been shown that the value of $\rho$ varies around 0.02 for parameters usually studied in DHS (Verma et al., 1980; Verma, 1991). A value of $b$ of around 30 households is often considered as adequate.

One possibility to decrease the scourge of design effect is to reduce the value of $b$ and hence increase, for a given $n$, the number of PSUs to be selected. Cost implications of increasing number of PSUs relative to units that are to be selected in a PSU (b) is important in deciding the optimum value of $b$. Purely from the point of view of precision of an estimate, it is generally true that the value '$a$' should be as large as possible. However, increasing number of PSUs affects costs of travel as well as time, whereas cost of increasing $b$ mainly refers to interviewing cost, although it is not directly proportional to the value of $b$. There are few field management issues as well as other survey requirements that also need to be given due consideration, for example, number of days that a data collection team stays in a PSU. It is often a requirement, to control non-sampling errors, to spend at least 2 days in a PSU (so that in first visit if a respondent is not available, she/he can be contacted again another day). The composition of field team is also important in this context (Roy and Pandey, 2008). Often a field team consists of more than one investigator, apart from field editors and supervisors. It might become cost-ineffective if a team travels all the distance to a PSU to interview only few respondents. Generally, interest in a survey is not to have any estimation at a PSU level, in household-based survey in rural India, where villages are a likely PSU, it might pay to have a reasonably large value of $b$ around 25–30. This would help reduce cost, provide a workable estimate at the village level and would not reduce the effective sample size much because of a low value of $\rho$ for parameters studied in many demographic surveys. All these tend to suggest that a drastic reduction in $b$ may not provide an optimum cluster size in every situation (for further discussion on the optimum cluster size, see Hansen et al., 1953). An example is given below to show the computation of estimated mean, sampling variance, design effect and intra-class correlation in the case of a sample selected in two stages.

**EXAMPLE 3.2**

In a contraceptive prevalence survey, in a hypothetical population, data on contraceptive use ($y = 1$ denoting use and 0 for non-use) of 720 women of age 15–44 were available. The women were selected in two stages. First 24 PSUs ($a$ = 24) were selected randomly and then from each selected PSU, 30 women ($b$ = 30) were selected again randomly. The idea is to estimate proportion of women using contraception in the population and to obtain sampling variance of the estimate, design effect and $\rho$ (fpc can be ignored). The data look like as below:

| S. No. of PSU | Number currently using contraceptives |
|---|---|
| 1 | 11 |
| 2 | 10 |
| 3 | 11 |
| 4 | 9 |
| 5 | 10 |
| 6 | 9 |
| 7 | 14 |
| 8 | 22 |
| 9 | 23 |
| 10 | 22 |
| 11 | 19 |
| 12 | 20 |
| 13 | 22 |
| 14 | 22 |
| 15 | 25 |
| 16 | 16 |
| 17 | 18 |
| 18 | 16 |
| 19 | 17 |
| 20 | 19 |
| 21 | 18 |
| 22 | 15 |
| 23 | 15 |
| 24 | 18 |

$y$ (total number of users) = 401; proportion using contraception $= \dfrac{401}{24 \times 30} = 0.5569$

$$\sum_{i=1}^{24} y_i^2 = (11^2 + 10^2 + \cdots + 18^2) = 7235$$

Using equation 3.10a (assuming selection to be without replacement)

$$v(\bar{y}) = \frac{1}{24.23.900}\left(7235 - \frac{401^2}{24}\right) = 0.0011.$$

$$\text{Hence, SE } (\bar{y}) = \sqrt{0.0011} = 0.033.$$

For a sample proportion $p$, estimated from a simple random sample, we know

$$v_{\text{SRS}}(\bar{y}) = \frac{pq}{n-1} = \frac{0.5569 \times 0.4431}{719} = 0.00034.$$

$$\text{Therefore, deff} = \frac{0.0011}{0.00034} = 3.2$$

This means the sampling variance of $y$ for the two-stage design employed here is 3.2 times higher compared with its sampling variance if the sample of 720 were selected using SRS.

Since there is no other factor in the design to alter the sampling variance, using equation 3.12,

$$\rho = \frac{\text{deff} - 1}{b - 1} = 0.076.$$

It is possible to compute the contribution of the second-stage sampling to the overall variance of 0.0011. From equation 3.6, one can estimate the value of $s_b^2$.

$$s_b^2 = \frac{\sum_{ij}(y_{ij} - \bar{y}_i)^2}{a(b-1)} = \frac{1}{a(b-1)}\left[\sum_i \left\{\left(y_{i1}^2 + y_{i2}^2 + \cdots + y_{ib}^2\right) - \frac{y_i^2}{b}\right\}\right]$$

In the first PSU, the value of $y$ is 11. This means, 11 women were users of contraceptive $(y_{1j} = 1)$ and remaining 19 were non-users $(y_{1j} = 0)$. Therefore,

$$s_b^2 = \frac{1}{24 \times 29}\left[\left(11 - \frac{11^2}{30}\right) + \left(10 - \frac{10^2}{30}\right) + \cdots + \left(18 - \frac{18^2}{30}\right)\right] = \frac{1}{696} \times 159.83339 = 0.22965$$

$$\text{The variance due to second-stage sampling} = \frac{0.22965}{24.30} = 0.00032$$

This means 29 per cent $\left(= \dfrac{0.00032}{0.0011}\right)$ of the variance is due to the second stage of sampling.

```
/*Example: 3.2 In a contraceptive prevalence survey, in a hypothetical population,
data on contraceptive use (y = 1 denoting use and 0 for non use) of 720 women of
age 15-44 were available. The women were selected in two stages. First 24 PSUs
(a = 24) were selected randomly and then from each selected PSU 30 women (b = 30)
were selected again randomly. The idea is to estimate proportion of women using
contraception in the population and to obtain sampling variance of the estimate,
design effect and ρ (fpc can be ignored).
The data at PSU level look like as below:

     PSU   Number currently       PSU   Number currently
           using contraceptives         using contraceptives
      1         11                 12        20
      2         10                 13        22
      3         11                 14        22
      4          9                 15        25
      5         10                 16        16
      6          9                 17        18
      7         14                 18        16
      8         22                 19        17
      9         23                 20        19
     10         22                 21        18
     11         19                 22        15
                                   23        15

However, in STATA we will use woman level data for 720 women of which 401 using
contraception and rest not using*/

/*STATA codes for calculating variance if there was no clustering and 720 women
were selected purely randomly using SRSWR*/
*The following code tells STATA that the design is SRSWR
svyset, clear  /*clearing any specification from before*/
svyset _n      /*data drawn using SRS*/
svy: mean contrause /*calculate mean and variance*/

[TRUNCATED OUTPUT]
svy: mean contrause /*calculate mean and variance*/
(running mean on estimation sample)

Survey: Mean estimation

Number of strata =       1          Number of obs     =      720
Number of PSUs   =     720          Population size   =      720
                                    Design df         =      719

--------------------------------------------------------------
             |              Linearized
             |        Mean   Std. Err.     [95% Conf. Interval]
-------------+------------------------------------------------
```

```
  contrause |   .5569444   .0185255      .5205739     .593315
--------------------------------------------------------------

Note: the mean is 0.5569444, and variance is 0.00034319 (0.0185255^2)

/*STATA codes for calculating variance if the sample was selected at 2 stages,
first selecting 24 PSU and then
within selected PSU 30 women selected randomly, no stratification was used*/

svyset, clear /*clearing any survey specification from before*/
svyset PSU || _n /*sample selected at 2 stages, first PSU and then with PSU
randomly*/
svy: mean contrause /*calculate mean and variance*/
estat eff, deff /*calculate design effect*/

[TRUNCATED OUTPUT]
. svy: mean contrause /*calculate mean and variance*/
(running mean on estimation sample)

Survey: Mean estimation

Number of strata =       1          Number of obs     =      720
Number of PSUs   =      24          Population size   =      720
                                    Design df         =       23

--------------------------------------------------------------
             |             Linearized
             |       Mean   Std. Err.     [95% Conf. Interval]
-------------+------------------------------------------------
   contrause |   .5569444   .0328148      .4890619    .6248269
--------------------------------------------------------------

Note: the mean is the same (0.5569444), and variance has increased to 0.00107681
(0.0328148^2)

. estat eff, deff /*calculate design effect*/

----------------------------------------------
             |             Linearized
             |       Mean   Std. Err.     DEFF
-------------+--------------------------------
   contrause |   .5569444   .0328148     3.1376
----------------------------------------------

Note: The design effect is 3.1376 (0.00107681/0.00034319)

**Calculating contribution of second stage sampling**

**In order to calculate contribution of second stage sampling we need to first
calculate 'within PSU variance'
**we run a one-way annova
loneway contrause PSU  /*runs one-way ANOVA*/
return list /*shows all scalar values saved*/
gen sd_w = r(sd_w) /*stores within psu sd*/
```

```
**now calculate variance due to second stage = Within PSU variance/Number of
clusters*Number of woman per cluster

gen var_2Stage = (sd_w)^2/(24*30)

display "2nd stage variance =  "var_2Stage

2nd stage variance =  .00031895

**calculate contribution of 2nd stage variance to overall variance
display "% contribution of 2nd stage variance to overall variance = "
(var_2Stage/0.00107681)*100 "%"

% contribution of 2nd stage variance to overall variance =  29.6201%
```

## 3.3 TWO-STAGE DESIGN WITH UNEQUAL CLUSTER SIZE

All the discussions in Section 3.2 pertained to situations where each cluster was assumed to have a fixed number of units ($B$) in a population. However, in practice, clusters are, generally, of unequal size which can pose few problems. Suppose a population consists of $A$ clusters, with $i$th cluster having $B_i$ units. That is,

$$N = B_1 + B_2 + \cdots + B_i + \cdots + B_A$$

Suppose target sample size (sample size that is desired to be selected) is $n$ and it is decided to select units randomly in both the stages. Considering

$$f = \frac{n}{N}$$

if,

$$f_1 = \frac{a}{A}$$

then,

$$f_2 = \frac{f}{f_1} = \frac{nA}{Na} = k$$

where $k$ is a constant. It is seen that probability of selecting any unit is

$$f_1 \times f_2 = f$$

which is also a constant. That is, the design will be self-weighting. However, number of units that actually gets selected is likely to differ from $n$. For example, number of units to be selected from $i$th PSU is equal to $k \times B_i$ (since $f_2$ is the probability of selecting a unit within a PSU). Since $B_i$ varies over PSUs, samples selected from a PSU will also vary depending on the size of a particular PSU that is selected. Larger the size of a selected PSU, higher will be the number of units selected from it. This variable sample size makes a few complications in the survey. It makes the field management difficult. Decisions about the size of a field team, number of days a field team would require to complete a PSU, etc. also become harder. It also complicates the analysis.

### 3.3.1 Estimation of Mean and Sampling Variance

Since sample size is not fixed, the estimate, instead of equation 3.4, becomes a ratio estimate.

The ratio estimate of $y$ can be expressed as[1]:

$$r = \frac{y}{x} = \frac{\sum_{i=1}^{a} y_i}{\sum_{i=1}^{a} x_i} \tag{3.14}$$

$x_i$ is the size of cluster $i$,

and its sampling variance is estimated as

$$v(r) = \frac{1}{x^2}[v(y) + r^2\, v(x) - 2r\, \mathrm{cov}(yx)] \tag{3.15}$$

where

$$v(y) = (1-f)\frac{a}{(a-1)}\left(\sum_{i=1}^{a} y_i^2 - \frac{y^2}{a}\right) \tag{3.16}$$

$$v(x) = (1-f)\frac{a}{(a-1)}\left(\sum_{i=1}^{a} x_i^2 - \frac{x^2}{a}\right) \tag{3.17}$$

and

$$\mathrm{cov}(yx) = (1-f)\frac{a}{(a-1)}\left(\sum_{i=1}^{a} y_i x_i - \frac{yx}{a}\right) \tag{3.18}$$

It can be noted that equations 3.16–3.18 are the usual formula adopted for the estimation of variance and covariance. They can also be derived from equation 3.10a.

It has been shown that both a ratio estimate and its estimated sampling variance are approximations (see section 2.6.3.2). The ratio estimate $r$ is not an unbiased estimate of the population value $R$ (which is in fact the population mean). However, the extent of bias can be negligible and considered a fairly good approximation for a properly designed survey (e.g., Kish, 1995). We illustrate the procedure of computing a ratio estimate and its sampling variance in the example presented below.

---

[1] The ratio estimate, based on '$a$' PSUs, can be computed in two ways: i) ratios can be computed for each PSUs and then they are suitably weighted to get the overall ratio, ii) one single ratio can be obtained (as in equation 3.14) based on the sum of all $y$ and $x$. It has been suggested that the second option is better, since ratios for individual PSUs are based on small numbers and hence show greater variability (Hansen et al., 1953).

## Example 3.3

In a household survey aimed to estimate prevalence of institutional delivery in a hypothetical population, 450 women, who had a birth in the last 1 year, were selected. A two-stage selection of 14 PSUs by PPS and selection at second stage in a manner to keep the design self-weighting was adopted. Following data show number of women selected ($x$) and number who delivered in an institution ($y$).

| S. No. | 1 | 2 | 3 | 4 | 5 | 6 | 7 | 8 | 9 | 10 | 11 | 12 | 13 | 14 |
|---|---|---|---|---|---|---|---|---|---|---|---|---|---|---|
| $y$ | 12 | 14 | 10 | 9 | 32 | 35 | 28 | 31 | 30 | 24 | 23 | 25 | 22 | 23 |
| $x$ | 25 | 25 | 24 | 23 | 42 | 40 | 40 | 41 | 39 | 31 | 29 | 30 | 30 | 31 |

Let us estimate the proportion of institutional delivery in the population and its sampling variance.

Since cluster size is variable, we need to use ratio estimate of $y$, that is,

$$r = \frac{\sum_i y}{\sum_i x} = \frac{318}{450} = 0.7067; \sum_i y_i^2 = (12^2 + 14^2 + \cdots + 23^2) = 8158,$$

$$v(y) = \frac{14}{13}\left(8158 - \frac{318^2}{14}\right) = 1006.7692; \sum_i x_i^2 = (25^2 + 25^2 + \cdots + 31^2) = 15084$$

$$v(x) = \frac{14}{13}\left(15084 - \frac{450^2}{14}\right) = 667.3846; \sum_i y_i x_i = (12.25 + 14.25 + \cdots + 23.31) = 10936$$

$$\text{cov}\,(yx) = \frac{14}{13}\left(10936 - \frac{318.450}{14}\right) = 769.5384.$$

Therefore, using equation 3.15

$$v(r) = \frac{1}{450^2}[1006.7692 + r^2 \times 667.3846 - 2r \times 769.5384] = 0.00125; \ \text{SE}(r) = 0.035$$

```
/*
Example 3.3: calculation of variance [STATA CODE]
Example 3.3 In a household survey aimed to estimate prevalence of institutional
delivery in a hypothetical population,450 women, who had a birth in the last one
year, were selected. A two stage selection of 14 PSUs by PPS and election at
second stage in a manner to keep the design self-weighting was adopted.
Following data show number of women selected (x) and number who delivered in an
institution (y).

PSU  1    2    3    4    5    6    7    8    9    10   11   12   13   14
Y    12   14   10   9    32   35   28   31   30   24   23   25   22   23
X    25   25   24   23   42   40   40   41   39   31   29   30   30   31
```

```
Problem: Calculate the ratio estimate and its variance*/

*******Using PSU level abridged data**********************

First, set up the data
Use STATA data editor to enter the above data (alternatively cut and paste (transpose)
from MS Excel into the data editor. Three variables will be created - PSU, x and
y with 14 observation to each*/

/*first calculate x^2 and y^2, xy*/

gen x_sq = x*x
gen y_sq = y*y
gen product_xy = x*y

/*store sum of x and y*/

sum x
gen sum_of_x = r(sum)

sum y
gen sum_of_y = r(sum)

sum x_sq
gen sum_of_x_sq = r(sum)

sum y_sq
gen sum_of_y_sq = r(sum)

sum product_xy
gen sum_of_product_xy = r(sum)

gen a = r(N)  /*number of clusters*/

/*calculate variances of v(x): v(sum_of_x), v(y):v(sum_of_y)*/

gen var_x = (a/(a-1))*(sum_of_x_sq - ((sum_of_x)^2/a))
gen var_y = (a/(a-1))*(sum_of_y_sq - ((sum_of_y)^2/a))

/*covariance of x and y  cov(yx)*/

gen cov_xy = (a/(a-1))*(sum_of_product_xy - (sum_of_x*sum_of_y)/a)

/*ratio estimate*/

gen r = sum_of_y / sum_of_x

**variance of r v(r) = 1/(sum_of_x)^2  [V(y) + r^2 V(x) - 2 r Cov(yx)]

gen var_r = (1/sum_of_x^2)*(var_y + r^2*var_x -2*r*cov_xy)

gen se_r = sqrt(var_r)

display as text "r=" r "  /////   " "SE(r)=" se_r

*****************using woman level data***********************

/*the calculation can be done much more easily if the data are available at the
woman level, with y taking value 1 if the woman had an institutional delivery
```

```
if 0 if not. The following codes can be used to calculate mean and variance of
y while the sample was done at two stages*/

svyset, clear /*clearing any survey specification from before*/
svyset psu || _n /*sample selected at 2 stages, first PSU and then within PSU
randomly*/
svy: mean y /*calculate mean and variance*/

[OUTPUT]
Survey: Mean estimation

Number of strata =         1           Number of obs      =       450
Number of PSUs   =        14           Population size    =       450
                                       Design df          =        13

-------------------------------------------------------------------
             |             Linearized
             |       Mean   Std. Err.       [95% Conf. Interval]
-------------+-----------------------------------------------------
           y |   .7066667   .0353069        .6303907     .7829426
-------------------------------------------------------------------
```

Before discussing what is meant by a properly designed survey mentioned above, let us examine other alternative designs. The design described above is a self-weighting design, but sample size selected in each selected PSUs are unequal. An alternative design is to select equal sample size from each PSUs selected in first stage. That is, select $b(=\frac{n}{a})$ elements from each of '*a*' PSUs selected in first stage. This will obviate the problems of variations in the cluster size. However, this design will not remain self-weighting. The probability of selecting a unit within *i*th PSU is equal to $\frac{b}{B_i}$ and this will vary with size of a PSU that is selected. Let us illustrate this with a small example.

### EXAMPLE 3.4

A hypothetical population consisting 1800 individuals is distributed in three clusters as:

| Cluster | A | B | C | Total |
|---|---|---|---|---|
| No. of persons | 300 | 600 | 900 | 1800 |

Suppose, we want to select a sample of 36 individuals from this population selecting two clusters.

Suppose that the two clusters are selected with EPSEM $\left(f_1 = \frac{2}{3}\right)$ and 18 individuals are targeted to be selected in each selected cluster. In the table below, following values are shown for each clusters under the two methods described above – one selecting variable number of units per cluster (i.e., in proportion of number of units within cluster) and second selecting equal number of units from each selected cluster. We calculate the following: i) probability of selecting a unit in a cluster ($f_2$), ii) expected number of units that would be selected if a cluster is selected and iii) overall probability of selecting a unit from a cluster.

| | Method 1 | | | Method 2 | | |
|---|---|---|---|---|---|---|
| | Cluster | | | Cluster | | |
| | A | B | C | A | B | C |
| | $f_2 = \frac{f}{f_1} = \frac{36}{1800} \div \frac{2}{3} = 0.03$ | | | $\frac{18}{300}$ | $\frac{18}{600}$ | $\frac{18}{900}$ |
| No. to be selected | $f_2 \times 300 = 9$ | $f_2 \times 600 = 18$ | $f_2 \times 900 = 27$ | 18 | 18 | 18 |
| Overall problem of selecting a unit | $\frac{2}{3} \times \frac{9}{300} = 0.02$ | $\frac{2}{3} \times \frac{18}{600} = 0.02$ | $\frac{2}{3} \times \frac{27}{900} = 0.02$ | $\frac{2}{3} \times \frac{18}{300} = 0.04$ | $\frac{2}{3} \times \frac{18}{600} = 0.02$ | $\frac{2}{3} \times \frac{18}{900} = 0.013$ |

It can be seen that in the first method, which keeps the design self-weighting, by deciding $f_2$ using the relation $f = f_1 \times f_2$, the overall probability of selecting a unit is same, equal to 0.02 but the disadvantage is that the sample size may vary; for example, if clusters A and B are selected, the size would be 27, with B and C it would be 45, etc. In the second method, equal number of units is selected from each cluster, so the sample size is fixed. But it is not self-weighting, for example, probability of selecting a unit from C, which has the largest size, is one-third of the probability that a unit belonging to A has.

### 3.3.2 Two desirable properties of the design

It is desirable for a multistage design to satisfy the following two conditions:

1) The design should be self-weighting, that is, units in a population are selected with EPSEM and 2) the number of units selected within a PSU, should be similar.

A design that is self-weighting is usually preferred. Although self-weighting design is not a panacea, it is a desirable property. It helps in getting an unbiased estimate of parameters of a population. One need not use weights to obtain an estimate. It simplifies the analysis. Although cost of computation may not be an advantage now, because of the presence of fast computing facilities, it does facilitate analysis, particularly in large-scale surveys. In a large-scale survey that requires analysis of a variety of variables at different levels (e.g., at individual, household or village level), consideration of appropriate weights require particular skills. Subclass variables often assume importance in analysis of a survey data. For example, in a survey on disease prevalence, apart from the prevalence itself, one may be interested in analysing prevalence by certain subclass variables, for example, age, education, etc. If a design is not self-weighting, one needs to be extra careful about the validity of these estimates.

The second desirable property, that is, the variation in cluster size should be minimum is important too, particularly since the sample estimate, in the presence of variations in cluster size, as shown in equations 3.14 and 3.15, are approximations and for validity of these approximations, particularly the sampling variance of the ratio, it is necessary to have low variations in cluster size.

### 3.3.3 Guidelines for Attaining Desired Property

It is clear that if PSU sizes are equal, it is easy to satisfy both the properties. A self-weighting design, in which PSUs are selected with EPSEM automatically leads to a fixed number of units selected at the second stage, that is, it will have same cluster size. The problem arises when PSU sizes are not same, as is generally the case. However, even in this case, it is possible to make a design self-weighting as well as have equal cluster size. It must be mentioned that although the two properties are desirable, it is not necessary that they need to be followed exactly, that is, if situation warrants, the cluster size may vary but not beyond a tolerable limit. Similarly, the attempt should be to make a design self-weighting, but it may deviate from EPSEM if needed. The bias arising due to the departure can be reduced if estimates are weighted properly. It should, however, be noted that larger the departure from an EPSEM design (or larger the variations among the weights), greater the sampling variance of an estimate (equation 2.6).

Two guidelines are presented that may help in visualising the extent of variations in the cluster size that can be tolerated and how to exercise control over such variations.

#### Guideline 1

For a given sample size, effort should be made to have as many PSUs as the resources would permit. It has been mentioned elsewhere that the bias in a ratio estimate decreases and becomes negligible with increase in number of PSUs (Hansen et al., 1953). For valid use of the ratio estimates, it is necessary to reduce the variations in cluster size $x_i$.

The coefficient of variation of $x$, denoted by $c(x)$, is obtained as

$$c(x) = \frac{\sqrt{v(\bar{x})}}{\bar{x}} = \frac{\sqrt{v\,(x)}}{x}$$

where $v(x)$ can be estimated from equation 3.17 and

$$\bar{x} = \frac{x}{a}$$

where $x$ denotes the total of $x$ values $\left(\sum x_i\right)$ and '$a$' is the number of PSUs selected. After simplification $c(x)$ can be written as

$$c(x) = \sqrt{\frac{(1-f)}{a}}\,\frac{s_x}{\bar{x}} \tag{3.19}$$

From the above equation, it can be noted that $c(x)$ depends on '$a$' and larger the value of $a$, that is, larger the number of PSUs selected, smaller is the coefficient of variation and, hence greater is the credibility of the ratio estimate. $c(x)$ can be alternatively calculated as

$$c(x) = \frac{1}{x}\sqrt{(1-f_a)\,a s_x^2} \tag{3.20}$$

where $f_a$ is equal to $\frac{a}{A}$. The suggested rule of thumb is that a value of $c(x)$ less than 0.1 is good, that is, the amount of bias in the estimate would be negligible and sampling variance of the estimate (equation 3.15) would also be a good approximation of the corresponding population estimate. According to Kish (1995), a value of $c(x)$ less than 0.2 can be tolerated. Bias in the ratio estimate also depends on the correlation between $y$ and $x$; lesser the magnitude of this correlation, lower will be the amount of bias.

Considering the example 3.3, one can compute the value of $c(x)$ as,

$$c(x) = \frac{\sqrt{667.3846}}{450} = 0.0574,$$ a value much within recommended limits for $c(x)$.

For a given sample size, increasing the number of PSUs, helps in reducing the sampling variance of $r$ (the ratio estimate), as evident from equation 3.15. It also, lends support to make the estimate a good approximation of the true variance. In addition, larger the number of PSUs to be selected, greater will be the scope for its stratification, which can also help in enhancing the precision of an estimate. Issues on stratification will be discussed in Section 3.4.

### Choice of PSU

The above discussion amply demonstrates the importance of PSUs in a multistage design. The choice of what should be an ideal sampling unit for consideration of PSU depends on a number of issues. The PSUs should include each and every unit in a population and its size (in terms of number of PSUs) should be as large as possible (equation 3.9). Choice of PSUs depends largely on what information is available at a particular level. Consider the case of India. In most of the household-based surveys in India, if estimates are to be obtained only at the level of, say, rural areas of the country, the choices could be states, districts, taluks/blocks and villages. One can easily avoid state as a choice. There are only 35 states, they are too large in terms of the population size. Apart from the small number of states, which is important as discussed above, selecting a fraction of the states and hence restricting the selection from them is not likely to provide a representative idea about the whole country. Every state (being a large unit) can have its own distinct characteristics. Consideration of village (the country has more than 6,00,000 villages) will have a wide spread of the sample and will tend to increase the cost substantially. Alternatively, one can consider either blocks (there are approximately 1,00,000 of tehsils) or even districts (total number of districts as per 2011 census is 647) as PSU. Another point that also merits consideration, in this context, is the availability of relevant characteristics of a PSU for their stratification. As the size of PSU increases, the chances of having the information also increase. For example, at state level, we have a large number of indicators that can be fruitfully used for stratification. This is also true, to a large extent for districts, but the availability reduces at village level. For some local level survey, if one wants to bifurcate villages into natural segments to consider as PSUs, no fruitful information on such units may be available for the purpose of stratification. Another issue to be considered is about the degree of clustering between units within a cluster. As the size of PSUs decreases, the extent of clustering tends to increase. One should attempt to keep the clustering effect as low as possible.

**Guideline 2**

Apart from an appropriate decision on PSUs, in a design, an attempt should be made to control the variations in cluster size. To recapitulate, the variations occur because in an EPSEM selection (mentioned as method 1 above), the number of units to be chosen in a selected PSU depends directly on its size. There are three distinct ways to control these variations in PSU size that would be discussed (Kish, 1995):

1. To control the size of units in a population that is considered as PSU.
2. To change selection procedure so that the twin purpose of having an EPSEM design as well as controlling the size of ultimate clusters would be possible.
3. Use stratification of PSUs suitably so that the number of units selected in PSUs within a stratum would not differ widely.

## 3.3.4 Ways to control variations in cluster size

### 3.3.4.1 Controlling size of clusters

One simple way, before PSUs are selected, is to reorganize them, to the extent possible, so that their size would not be very different from each other. For example, if villages are considered as PSUs, small villages can be merged suitably with other nearby villages and considered as one sampling unit for the purpose of PSU selection as well as for selection of second-stage units, if it gets selected. Similarly, larger size villages, if possible, can be divided suitably. Larger size villages, generally, would have clear administrative or natural divisions as well as estimate of their size for further segmentation discussed in chapters 4, 5 and 10.

### 3.3.4.2 Alternative selection procedure

The selection procedure can be changed suitably to remedy the situation. If information on PSU size is available, one can utilize this information and adjust the probability of their selection accordingly. In other words, PSUs of varying size can be selected with probability proportional to their size (PPS). In the second stage, a fixed number of units can be selected giving equal chance of selection to each unit from a selected PSU. Let us consider the example 3.4 to illustrate the procedure. We want to select two clusters. The probability of selecting an individual in cluster A can be obtained as the product of following two probabilities:

$$\text{the probability of selecting cluster A in first stage } (f_1) = \frac{2 \times 300}{1800}$$

The denominator is the total population size, and in the numerator, 2 refers to the number of clusters being selected and 300 is the size of A. If selected, 18 units will be selected giving equal chance to each of the 300 units in it.

$$\text{The probability of selecting an individual unit in cluster A } (f_2) = \frac{18}{300}$$

Hence, the overall probability of selecting an individual from A is

$$= \frac{2 \times 300}{1800} \times \frac{18}{300} = \frac{36}{1800}$$

Similarly, probability of selecting an individual from cluster C is

$$= \frac{2 \times 900}{1800} \times \frac{18}{900} = \frac{36}{1800}$$

Therefore, in a two-stage design, if PSUs are selected with PPS sampling and then a same number of units are selected with equal probability, the overall probability of selecting a unit, irrespective of which PSU it belongs, will be a constant and equal to the overall sampling fraction ($f$). It can be checked that

$$f = f_1 \times f_2 = \frac{36}{1800}$$

Let $N$ denotes the overall population size and $B_i$ denotes the size of $i$th PSU. That is,

$$N = \sum_1^A B_i$$

where $A$ is the number of PSUs in the population. If '$a$' is the number of PSUs to be selected and $b$ is the number of second-stage units selected in a selected PSU, then the sample size, $n$ is given by

$$n = a \times b$$

The probability of selecting an individual unit in any selected PSU can be obtained by calculating,

$$\text{First-stage probability of selection} = f_1 = \frac{a \times B_i}{B} \tag{3.21}$$

and

$$\text{Second-stage probability of selection} = f_2 = \frac{b}{B_i} \tag{3.22}$$

Finally, overall selection probability of an individual unit = $f = \frac{a \times B_i}{B} \times \frac{b}{B_i} = \frac{a \times b}{B} = \frac{n}{N}$, which is a constant.

Hence, selection in this way would ensure that each and every unit in the population has equal chance of being included in the sample, that is, it would be a self-weighting design. In addition, since an equal number of individual units are selected from all selected PSUs, there is no variation in the cluster size (except for some possible non-response, which will be discussed later). This, therefore, satisfies both the conditions and hence can be a very potent option in a multistage design.

#### Case of a three-stage design

Suppose a three-stage design is to be adopted where $k$ number of blocks are selected from the $i$th selected PSU, and then from each of the $k$ blocks equal number of units $\frac{b}{k}$ are selected so that $b$ units are selected from each of the '$a$' selected PSUs. Let $C_{ij}$ be the size of the $j$th block in $i$th PSU. It can be seen that

$$B_i = \sum_j C_{ij}$$

If blocks in the second stage are also selected with PPS and $\frac{b}{k}$ individual units are selected in the third stage with equal probability, then probability of selecting an individual unit from $j$th block of $i$th PSU will be ($f$)

$$f = f_1 \times f_2 \times f_3$$

where

$$f_1 = \text{probability of selecting a PSU} = \frac{a B_i}{B}$$

$$f_2 = \text{probability of selecting } j\text{th block in } i\text{th PSU} = \frac{k C_{ij}}{B_i}$$

$$f_3 = \text{probability of selecting an individual in } j\text{th block of } i\text{th PSU} = \frac{\frac{b}{k}}{C_{ij}}$$

It can be easily seen that

$$f = \frac{a \times b}{B} = \frac{n}{N}$$

Since PSUs vary in size, it seems quite logical to give higher chance to a PSU having larger number of units. It is likely that large sampling units will be relatively more heterogeneous than a sampling unit that is smaller in size. In Indian rural situation, for example, it has been generally observed that compared to average sized villages, proportion of illiterate and other socially disadvantaged population is higher in small size villages. It, therefore, makes perfect sense to give homogeneous units less chance for selection and capture more variability in the selected sample. Selection of equal number of units at the last stage, however, helps make such a design self-weighting. Selection of more units from large sized PSUs would disturb both, without any adequate gain (as already mentioned, the contribution of within-cluster variation to the sampling variance is usually small).
Such designs are widely used, particularly in household-based surveys. They are popularly used in DHS conducted periodically in many countries in the developing regions. It is, therefore, discussed in further details.

#### *3.3.4.2.1 Alternatives when information on size of PSUs refers to a past period*

In order to select PSUs using PPS sampling, one needs to have up to date information on size of PSUs. Such data often come from the *census*, conducted periodically. It means that updated data on size of the PSUs may not be available at the time of conducting a survey during the intercensal period. For example, available data might refer to time '$t$' for a survey to be conducted at time '$t + k$'.

Ignoring the time gap, one can utilize equations 3.21 and 3.22 using available data for time $t$. However, it is obvious that size of village will change over time between $t$ and $t + k$, and these changes will depend on the gap in time (i.e., the value of $k$). If one uses the data at time $t$ for sampling, it would mean one is ignoring the changes. The additions (i.e., those who were not a member at $t$ but are so at $t + k$) will get zero chance of being included in the sample contrary to the basic principle of a probability sample, and those who cease to be a member of the population at time $t + k$, if selected would tend to reduce the sample size. In both the cases, particularly in the former, there is possibility of bias creeping in the design. This, however, is a very common problem in sampling design. We discuss two alternative solutions below.

Let us assume that the population under study consists of A PSUs, $H_i$ households in $i$th PSU and a total of $H$ households at time $t$ (e.g., census), and the survey is being conducted at time $t + k$.

### Solution 1

Under PPS, the probability of selecting a PSU ($f_1$) is based on $H_i$ s. That is,

$$f_1 = \text{probability of selecting } i\text{th PSU} = \frac{a \times H_i}{H}$$

After selection of PSUs, a household listing operation is carried out in all the selected PSUs. Let $H_i'$ denotes the number of households in the $i$th PSU at the time of listing ($t + k$). Then the required number of households ($b$) is selected in $i$th PSU giving equal probability ($f_2$) to each household within it. That is,

$$f_2 = \frac{b}{H_i'}$$

The overall probability of selecting a unit in $i$th PSU will, therefore, be:

$$f = f_1 \times f_2 = \frac{a \times H_i}{H} \times \frac{b}{H_i'} = \frac{ab}{H}\left(\frac{H_i}{H_i'}\right) \tag{3.23}$$

However, since this is not a constant, the probability of selection of a unit will not remain the same. It will vary by PSU although all units within a PSU will have equal chance of selection. The variation in the probability occurs because of the factor $\left(\frac{H_i}{H_i'}\right)$ which depends on the relative growth of different selected PSUs during the period between $t$ and $t + k$. Clearly, a PSU which grows faster will tend to have a lower probability of selection compared to the ones whose growth rates have been lower. If elapsed time ($k$) is not large, the variation in the probability is likely to be marginal and tolerable. The bias created by differential growth can be taken care of by using appropriate weights. The weight for units in $i$th selected PSU will thus be

$$\text{Weight for units in } i\text{th PSU}(w_i') = \frac{1}{\text{probability of selecting a unit in } i\text{th PSU}}$$
$$= \frac{H \times H_i'}{ab \times H_i}$$

It can be checked that the weight will be equal if there is no change in the population size of $i$th PSU or if the rate of growth is same for all PSUs $\left(\frac{H_i'}{H_i} \text{ is constant}\right)$. Within a PSU, if $b$ units are selected with equal chance, $w_i'$ will be same for all the $b$ units. It is instructive to check the sum of the weights.

$$\sum_{ij} w_{ij}' = \sum_{ij} \frac{H}{ab} \times \frac{H_i'}{H_i}$$
$$= \frac{H}{ab} \sum_i b \times \frac{H_i'}{H_i}$$
$$= \left( H \frac{1}{a} \sum_{i=1}^{a} \frac{H_i'}{H_i} \right)$$

The value in the parenthesis indicates the average growth of '$a$' PSUs during the period $k$. This can be considered as an estimate of the growth of A PSUs in the population $\left(= \frac{H'}{H}\right)$. Hence, sum of weights would give an estimate of the total number of units in the population at survey time (although purpose is not really to estimate the population total in this way and there are better approaches available for estimating population in the field of demography).

Weights can be used to make correction for bias in an estimate that occurs because of not selecting units with equal probability. They can be normalized so that they are calibrated around the sample size. If $w_i$ denote the normalized weight for $i$th PSU, then

$$\left( w_i = w_i' \frac{n}{\text{sum of all the } n \text{ weights}} \right) \tag{3.24}$$

where $n$ is the sample size, that is, the number of interviews actually conducted. Henceforth reference of weights will mean the normalized weights, unless otherwise stated.

Generally, there will be some non-response and the total number of interviews actually conducted will be lower than $n$. The non-response rate will also vary by PSU and can introduce a bias and hence would need corrections. The non-response rate is often found to be higher among educated individuals, working individuals and among those who stay in urban areas. Suppose the variable under study is to estimate the prevalence of institutional delivery which is positively correlated with education. A higher non-response leading to a greater loss of educated persons in the interviewed sample would tend to underestimate the prevalence. There is a need, therefore, to take care of the non-response and make suitable adjustment. Non-response adjustment can be done at individual PSU level. One can use, instead of equation 3.23, the probability of interviewing a person in $i$th PSU as

$$= \left(\frac{ab}{H}\right) \times \left(\frac{H_i'}{H}\right) \times (r_i)$$

where $r_i$ is the response rate observed in $i$th PSU. The other steps in the computation of weights will remain as mentioned before, except for the value of $n$ in equation 3.24 which will now be replaced by $n'$, the number of households actually interviewed.

Apart from the differential growth of the selected PSUs, non-response also contributes to make the probability of selection of a unit to deviate from equality. Larger the variations in probability of selection of a unit or its weight, greater will be the sampling variance.

It may be noted that the use of $y_i$ and $x_i$ in equations 3.14 to 3.18 assumes a self-weighting design ($w_i = 1$). If weights are used, the values of $y_i$ and $x_i$ need to be multiplied by appropriate weights. That is, values of $w_i y_i$ and $w_i x_i$ should be used in all the above equations, instead of $y_i$ and $x_i$.

The use of weights generally increases the variance of $x$ ($v(x)$), which, in turn, contributes to an increase in the variance of $r$ (equation 3.15) as well as $c(x)$ (equation 3.19). The $v(r)$ increases by a factor, known as multiplicative effect (Meff$_w$), where

$$\text{Meff}_w = \left(1 + \frac{s_w^2}{\bar{w}^2}\right) \tag{3.25}$$

$\bar{w}$ is the mean value of weights and $s_w^2$ is its standard deviation (Kish, 1995). In other words,

$$\text{Meff}_w = 1 + (\text{coefficient of variation of weights})^2$$

For a self-weighting design, which means each and every selected unit has same weight, equal to one, the value of Meff$_w$ is 1.

There is one disadvantage in making non-response adjustment at PSU level. Since number of individuals sampled in a PSU (b) is not sufficiently large, values of $r_i$ can fluctuate widely and will tend to increase variations in weights. To reduce the variations in weights due to non-response adjustment, the computation of non-response rate can be made by combining few PSUs belonging to a particular stratum, this will make the non-response rate relatively stable and reduce the value of Meff$_w$.

## Example 3.5

Let there be four villages, in a hypothetical population, consisting of 1000 households in 2001. A survey was conducted to estimate average daily wage in the population in 2006. Households were selected in two stages, in the first stage, two villages were selected by PPS using information about size in 2001 and then 10 households were selected with equal probability from each selected village after listing households in the two selected villages. Villages 1 and 3 were selected. There was one non-response in village 1 and three in village 3. Information on village size and daily wages reported by 16 selected and interviewed households are given below.

**(i)** Village size

| S. No. | Size of village | |
|---|---|---|
| | 2001 | 2006 (from listing) |
| 1 | 120 | 150 |
| 2 | 180 | |
| 3 | 500 | 750 |
| 4 | 200 | |
| Total | 1000 | |

**(ii)** Reported daily wages

| Village | 1 | 1 | 1 | 1 | 1 | 1 | 1 | 1 | 1 | 3 | 3 | 3 | 3 | 3 | 3 | 3 |
|---|---|---|---|---|---|---|---|---|---|---|---|---|---|---|---|---|
| S. No. of HHs | 1 | 2 | 3 | 4 | 5 | 6 | 7 | 8 | 9 | 10 | 11 | 12 | 13 | 14 | 15 | 16 |
| Wage | 25 | 40 | 60 | 10 | 30 | 80 | 20 | 40 | 35 | 120 | 150 | 90 | 300 | 75 | 100 | 200 |

Our problem is to estimate average daily wage in the population.

Using equation 3.23, probability of selecting a household in village

$1 = \frac{2 \times 120}{1000} \times \frac{10}{150} = 0.016000$

and probability of selecting a households in village $3 = \frac{2 \times 500}{1000} \times \frac{10}{750} = 0.01333.$

The calculation of weights to adjust for differential growth as well as non-response is illustrated below:

| Village (PSU) | S. No. households interviewed | Daily wages ($y$) | Weight (= 1/probability of its selection) ($w'_{ij}$) | Weight ($w''_{ij}$) adjusted for non-response $= \left(\frac{w'_{ij}}{R_i}\right)$ | Normalized weight ($w_{ij}$) = $w''_{ij} \times \left(\frac{16}{1375.018748}\right)$ |
|---|---|---|---|---|---|
| 1 | 1 | 25 | 62.500000 | 69.444444 | 0.808070 |
| 1 | 2 | 40 | 62.500000 | 69.444444 | 0.808070 |
| 1 | 3 | 60 | 62.500000 | 69.444444 | 0.808070 |
| 1 | 4 | 10 | 62.500000 | 69.444444 | 0.808070 |
| 1 | 5 | 30 | 62.500000 | 69.444444 | 0.808070 |
| 1 | 6 | 80 | 62.500000 | 69.444444 | 0.808070 |
| 1 | 7 | 20 | 62.500000 | 69.444444 | 0.808070 |

| Village (PSU) | S. No. households interviewed | Daily wages ($y$) | Weight (= 1/probability of its selection) ($w'_{ij}$) | Weight ($w''_{ij}$) adjusted for non-response $=\left(\frac{w'_{ij}}{R_i}\right)$ | Normalized weight ($w_{ij}$) = $w''_{ij}\times\left(\frac{16}{1375.018748}\right)$ |
|---|---|---|---|---|---|
| 1 | 8 | 40 | 62.500000 | 69.444444 | 0.808070 |
| 1 | 9 | 35 | 62.500000 | 69.444444 | 0.808070 |
| 3 | 10 | 120 | 75.001875 | 107.145536 | 1.246767 |
| 3 | 11 | 150 | 75.001875 | 107.145536 | 1.246767 |
| 3 | 12 | 90 | 75.001875 | 107.145536 | 1.246767 |
| 3 | 13 | 300 | 75.001875 | 107.145536 | 1.246767 |
| 3 | 14 | 75 | 75.001875 | 107.145536 | 1.246767 |
| 3 | 15 | 100 | 75.001875 | 107.145536 | 1.246767 |
| 3 | 16 | 200 | 75.001875 | 107.145536 | 1.246767 |

$R_i$ is the response rate of $i$th selected PSU.

$$\text{Sum of weights} = \sum_{ij} w'_{ij} = (9\times 69.444444\ +\ 7\times 107.145536)\ =\ 1375.018748$$

Estimated average daily wage of the population is

$$\frac{\sum_{ij} w_{ij} y_{ij}}{\sum_{ij} w_{ij}} = \frac{(25\times 0.80807 + 40\times 0.80807 + \cdots + 200\times 1.246767)}{16} = 97.8217$$

It can be checked that the value of the estimate will remain same if $y$ values are weighted by $w''_{ij}$ instead of the normalized weight, $w_{ij}$. However, if $w'_{ij}$, that is, weights without adjusting for non-response, which is also called base weights, is used instead of $w_{ij}$, the estimated daily wage will be 90.9202. This difference occurs because the daily wage is higher in village 3 compared to that in village 1. The non-response adjustments at the PSU level is expected to reduce the bias.

The value of $\text{Meff}_w$ is, however, as could be expected, lower in the case of $w'_{ij}$, which ignores non-response adjustments and hence will have less variations among weights. It can be checked that its value is 1.008326. $w''_{ij}$ and $w_{ij}$ have similar variations in weights. They are both equal to 1.047362.

## Solution 2

The solution discussed earlier allows the design to deviate from its self-weighting nature and thereby allowing the probability of selection of units to vary, albeit slightly. The present alternative helps in keeping the design self-weighting. In this case, the overall sampling fraction ($f$) or the probability of selecting a unit is estimated in the beginning. If $n$ is the target sample size and $H'$ is the estimated number of households in the area at survey time, $t+k$, then

$$f = \frac{n}{H'}$$

The estimation of $H'$ can be done with the help of an appropriate population projection technique. Selection of '$a$' PSUs can then be done as usual with PPS, that is,

$$f_1 = \text{ probability of selecting } i\text{th PSU} = \frac{a \times H_i}{H}$$

Assuming that it would be a self-weighting design, that is,

$$f = f_1 \times f_2$$

The value of $f_2$ can be estimated as

$$f_2 = \frac{f}{f_1} = \frac{n}{H'} \times \frac{H}{aH_i}$$

Therefore, the expected number of units to be selected from $i$th PSU will be

$$\begin{aligned} &= \left( \frac{n}{H'} \times \frac{H}{aH_i} \right) H_i' \\ &\left( \frac{n}{a} \right) \times \left( \frac{H_i'}{H_i} \right) \times \left( \frac{H}{H'} \right) \end{aligned} \tag{3.26}$$

where $H_i'$ is the number of units found (at the time of listing) in $i$th PSU at the time of survey. Though it would be a self-weighting design, the number of units to be selected in each PSU would vary. The extent of variation would depend on the growth rate of $i$th PSU relative to that of the total population (equation 3.26). The growth of $i$th PSU is expected to be equal to that for the entire area $\left( \text{i.e., } \frac{H_i'}{H_i} = \frac{H'}{H} \right)$ and hence the number of units expected to be selected in $i$th PSU is $\frac{n}{a} = b$.

But in some PSUs, particularly in urban areas, the growth can be excessively high leading to a much higher value of the number of units to be selected from these PSUs. Similarly, in some PSUs, such as in small villages, the growth can be less than the average and the number of units to be selected from it would be less than $b$. Such variations in the number of units selected would tend to increase $v(x)$ as well as $c(x)$, which, as mentioned, can have adverse effects on the design. If need arises, a restriction may be imposed to avoid excessive variations in the number of units selected in a PSU. For example, if $b\left(=\frac{n}{a}\right)$ is the expected value, it can be decided that it will not be allowed to be less than $\frac{b}{2}$ and greater than $2b$. The overall probability of selection of a unit, where such adjustment is made, will change accordingly. This will, however, make the design to deviate from remaining self-weighting. The variation in weights, as a result of such adjustment is likely to increase and so also the value of $\text{Meff}_w$. However, the value of $c(x)$ gets controlled accordingly. This trade-off between the values of $c(x)$ and $\text{Meff}_w$ are sometimes necessary. The method is further elaborated in chapter 10.

#### 3.3.4.3 Stratification of PSUs to reduce variations in cluster size

Stratification is the other procedure that can also help in reducing the variations in number of units selected within PSUs. Apart from this role, stratification is also advantageous in increasing the precision of estimates in a multistage design. The use of stratification is discussed in Section 3.4.

## 3.4 STRATIFICATION IN MULTISTAGE DESIGN

Stratification refers to grouping of units having a similar characteristic. If the characteristic considered is size of PSU, it helps reduce the variation in cluster size (i.e., the number of units selected in each cluster). If PPS is not used and sampling units in both the first and second stages are selected with equal chance to each sampling units, PSUs with larger size will result in greater number of units to be selected at the second stage and vice versa (discussed in example 3.4). If the list of PSUs is stratified by PSU size, variation in cluster size within stratum can be controlled to a great extent. If within a stratum, units are selected by EPSEM, the cluster size will be roughly equal. The procedure of estimating mean and variance, within a stratum will be the same as discussed before. Let $f_l$ denote the sampling fraction $\left(\frac{n_l}{N_l}\right)$ in $l$th stratum such that the probability of selection of $n_l$ units is same for all units. If $f_l$ is equal to $f$, the overall probability of selection, it would mean that a proportional allocation has been used for distribution of the overall sample to different strata and the overall design will be self-weighting. If a disproportionate allocation is adopted to distribute the overall sample to different strata, the overall design will not be self-weighting but within a stratum it will still remain close to self-weighting. In either case, therefore, the mean and variance estimated for each stratum can be suitably pooled (discussed in Section 2.3.2.1) to get the overall estimate.

In order to get EPSEM selection within a stratum, as discussed earlier, one can adopt either equal probability selection at different stages or select first stage units (also later stages, if considered more than two stages of sampling, except the last) by PPS and same number of units with equal chance at the second (or last stage).

Generally, stratification is used for selection of PSUs. This is recommended because the gain in stratification is higher in cluster sampling compared to that when units are sampled directly. This is because the gain is compared relative to the variance of a design. In the case of sampling of units directly, the amount of gain due to stratification is compared to $S^2$, the overall variance, whereas it is done with $S_a^2$ in a cluster sampling. Since $S_a^2$ is usually lower than $S^2$, particularly when $\rho$ is small, gain due to stratification happens to be higher in cluster than in element sampling (for details on comparative gain by stratification, see

Hansen et al., 1953; Kish and Anderson, 1978). Since the sampling variance in a multistage design depends more on PSU level variations, stratification of second stage (or subsequent stage units) units does not yield much gain.

Stratification of PSUs is particularly helpful and the principles of stratification discussed in Chapter 2 are applicable in this case. The number of variables that could be considered for stratification often gets limited and depends on the number of PSUs that are to be selected. A general requirement is to select at least two PSUs from a stratum.

Therefore, one needs to check and select diligently from among the available variables. For household-based surveys at state level, in rural India, for example, village is often a viable unit to consider as PSU. Census provides a number of information at village level, such as level of female literacy, per cent of scheduled caste and scheduled tribe population, per cent of males in non-agricultural activity, apart from its size and geographical region that can be used fruitfully for stratification in most of the surveys. One can examine relationship between these variables and the study variable to decide about their inclusion as a variable for stratification.

A variable, such as level of female literacy, which is found to have dominant relationship with most of the demographic variables, is often considered for the stratification. The PSUs are suitably ranked according to their level of female literacy before selecting them systematically (Goodman and Kish, 1950). This can be considered, for the purpose of estimation of the parameters, as selection of two units from each strata or selection of a pair from the ordered list (Kish, 1995, p. 193–195). Example 3.9 provides further illustration on it. Further discussion on stratification employed in one of the largest DHS, that is, NFHS, in India can be found in Roy and Pandey (2008).

The estimation of mean and variance in the case of stratified cluster sampling is presented below considering both unequal and equal cluster sizes.

### 3.4.1 Estimation of Parameters in Unequal Cluster Size

We discuss below the procedure of estimating mean and variance for two types of stratified sampling: 1) more than two clusters are drawn from each stratum and 2) when two clusters are drawn from each stratum.

**(i)** *More than two clusters selected in each stratum*

Let there be $L$ strata into which A clusters are divided in a population with $A_l$ clusters belonging to $l$th strata. We select a total of 'a' clusters in such a way that $a_l$ clusters are selected from $l$th strata and the ultimate cluster size in it is $b_{li}$.

In the case of unequal cluster size, the sample size will also be a variable and the estimate of population parameter will be a ratio estimate as

$$r = \frac{y}{x}. \tag{3.27}$$

Both the numerator and denominator are computed based on their respective sums (using appropriate weight for each unit). This can be obtained in two steps, by computing their sum in $l$th strata and then summing over all the strata. That is,

$$r = \frac{\sum_{i=1}^{L}(y_{l1} + y_{l2} + \cdots y_{li} + \cdots y_{la_l})}{\sum_{i=1}^{L}\left(x_{l1} + x_{l2} + \cdots x_{li} + x_{la_l}\right)}$$

where $y_{li}$ is the sum of $y$ values for all the $b_{li}$ units of $i$th PSU (i = 1, 2...$a_l$) in the $l$th stratum. Similarly, $x_{li}$ is the sample size for $i$th PSU in $l$th stratum or the sum of weights of the $b_{li}$ units, when weights are used. With symbols, the above procedure may appear little complicated. However, if we have a data file with y values and their weights in two separate columns, the computation in equation 3.27 becomes simple. It would be a ratio of sum of the product of $y$ and their weights divided by the sum of weights. If stratification is done considering PSU size, the variation in cluster size ($b_{li}$) within the stratum is expected to be marginal which would provide the necessary support for the ratio estimation. The sampling variance of $r$ can be computed as,

$$v(r) = \frac{1}{x^2}\left[\sum_l v(y_l) + r^2 \sum_l v\,(x_l) - 2r\sum_l \text{cov}\,(y_l x_l)\right] \tag{3.28}$$

The variance and covariance for each stratum can be estimated following in equations 3.16–3.18, that is,

$$v(y_l) = \frac{(1-f_l)a_l}{(a_l-1)}\left[\sum_{i=1}^{a_l} y_{li}^2 - \frac{y_l^2}{a_l}\right] \tag{3.29}$$

$$v(x_l) = \frac{(1-f_l)a_l}{(a_l-1)}\left[\sum_{i=1}^{a_l} x_{li}^2 - \frac{x_l^2}{a_l}\right] \tag{3.30}$$

$$v(y_l x_l) = \frac{(1-f_l)a_l}{(a_l-1)}\left[\sum_{i=1}^{a_l} y_{li} x_{li} - \frac{y_l x_l}{a_l}\right] \tag{3.31}$$

where $f_l = \frac{a_l}{A_l}$

Let us illustrate this with an example.

### EXAMPLE 3.6

Consider example 3.3 and assume that the PSUs in the population were stratified into three strata before selection. A total of 14 PSUs were selected proportionally from the three strata. PSUs 1–4, 5–9 and 10–14 were selected from stratums 1, 2 and 3, respectively. Estimates of variance (ignoring the values of fpc) are calculated as below using data provided in tables:

Data for stratum 1

| PSU | 1 | 2 | 3 | 4 |
|---|---|---|---|---|
| $y$ | 12 | 14 | 10 | 9 |
| $x$ | 25 | 25 | 24 | 23 |

Computation of variance of $r$:

$$v(y) = \frac{4}{3} \times \left[(12^2 + 14^2 + 10^2 + 9^2) - \frac{(12+14+10+9)^2}{4}\right] = 19.6667$$

$$v(x) = \frac{4}{3} \times \left[ (25^2 + 25^2 + 24^2 + 23^2) - \frac{(25+25+24+23)^2}{4} \right] = 3.6667$$

$$\text{cov}(yx) = \frac{4}{3} \times \left[ (12.25 + 14.25 + 10.24 + 9.23) - \frac{(12+14+10+9)(25+25+24+23)}{4} \right] = 7.6667$$

The values for the other two strata can be similarly computed, which are shown in the table below:

| | Stratum 1 | Stratum 2 | Stratum 3 | Total |
|---|---|---|---|---|
| $v(y)$ | 19.6667 | 33.5 | 6.5 | 59.6667 |
| $v(x)$ | 3.6667 | 6.5 | 3.5 | 13.6667 |
| $\text{cov}(yx)$ | 7.6667 | 3.25 | 0.75 | 11.6667 |

The estimate of $r$ remains same as in example 3.3, that is, 0.7067.

```
/*Example 3.6:
Please refer to example 3.3
Variance and covariance can be estimated for each stratum using STATA codes given
before when data are available abridged form for each PSU.

Here we will demonstrate how one can specify strata in STATA design command and
calculate mean and variance using woman level data described before.

We add a new variable to the dataset "strata" defining stratification discussed
in the example" */

svyset, clear /*clearing any survey specification from before*/
svyset psu, strata(strata) || _n /*sample selected at 2 stages, first PSU and
then with PSU randomly*/
                                 /*option strata defines stratified sampling at the
first stage*/
svy: mean y /*calculate mean and variance*/

[OUTPUT]
Survey: Mean estimation

Number of strata =       3           Number of obs      =       450
Number of PSUs   =      14           Population size    =       450
                                     Design df          =        11

--------------------------------------------------------------
             |             Linearized
             |       Mean   Std. Err.     [95% Conf. Interval]
-------------+------------------------------------------------
           y |   .7066667   .0157139      .6720806    .7412527
--------------------------------------------------------------

Note: Check number of strata: 3, SE reduced considerably to 0.0157139
```

$$v(r) = \frac{1}{450^2}[59.6667 + (0.7067^2) \times 13.6667 - 2 \times 0.7067 \times 11.6667] = 0.000247$$

$$\text{SE}(r) = \sqrt{0.00025} = 0.0157$$

**(i)** *Two clusters are selected per stratum*

In this case, the ratio is estimated as

$$r = \frac{\sum_{l=1}^{L}(y_{l1} + y_{l2})}{\sum_{l=1}^{L}(x_{l1} + x_{l2})} \tag{3.32}$$

Here, $L = \frac{a}{2}$

The numerator and the denominator above represent totals of $y$ and $x$ variables for the two clusters selected in $l$th strata. Its variance is calculated as:

$$v(r) = \frac{(1-f)}{x^2}\left[\sum_{l=1}^{L}(y_{l1} - y_{l2})^2 + r^2\sum_{l=1}^{L}(x_{l1} - x_{l2})^2 - 2r\sum_{l=1}^{L}(y_{l1} - y_{l2})(x_{l1} - x_{l2})\right] \tag{3.33}$$

Let us illustrate this with an example.

### Example 3.7

Consider example 3.6 and derive the estimates assuming that the 14 PSUs were selected systematically from an ordered list (i.e., as if two PSUs are selected in each of the seven strata).

To compute $v(r)$,

$$v(y) = \sum_{l=1}^{7}(y_{l1} - y_{l2})^2 = [(12-14)^2 + (10-9)^2 + \cdots + (22-23)^2] = 64$$

$$v(x) = \sum_{l=1}^{7}(x_{l1} - x_{l2})^2 = [(25-25)^2 + (24-23)^2 + \cdots + (30-31)^2 = 72$$

$$\text{cov}(yx) = \sum_{l=1}^{7}(y_{l1} - y_{l2})(x_{l1} - x_{l2}) = [(12-14)(25-25) + \cdots + (22-23)(30-31)] = 61$$

$$v(r) = \frac{1}{450^2} \times [64 + 0.7067^2 \times 72 - 2 \times 0.7067 \times 61] = 0.000068; \text{SE}(r) = \sqrt{0.000068} = 0.0082$$

### 3.4.2 Estimation of Parameters in Equal Cluster Size

We assume that units within a stratum are selected by EPSEM.

**(i)** *More than two clusters selected in each stratum*

Since the cluster size is equal, the estimate of mean and variance of $y$ can be obtained directly. If $b_l$ units are selected from each of $a_l$ PSUs, and $\bar{y}_l$ denotes the sample mean of $l$th stratum, then

$$\bar{y}_l = \frac{\sum_{i=1}^{a_l}\sum_{j=1}^{b_l} y_{lij}}{n_l} \tag{3.34}$$

where $n_l = a_l b_l$

The overall sample mean $(\bar{y})$ is obtained as

$$\bar{y} = \sum_{l}^{L} W_l \bar{y}_l \tag{3.35}$$

where $W_l$ denotes the proportion of the total population in $l$th stratum $\left(= \frac{N_l}{N}\right)$, and,

$$v(\bar{y}) = \sum_{l}^{L} W_l^2\, v(\bar{y}_l), \tag{3.36}$$

where,

$$v(\bar{y}_l) = (1 - f_l)\frac{1}{a_l}\left[\frac{1}{(a_l - 1)b_l^2}\left(\sum_{k=1}^{a_l} y_{lk}^2 - \frac{y_l^2}{a_l}\right)\right]. \tag{3.37}$$

## Example 3.8

Consider example 3.2 and let us assume that the PSUs in the population were stratified into three strata before selection. The PSUs 1–7, 8–15 and 16–24 were selected from stratums 1, 2 and 3, respectively. Assuming that the allocation is done proportional to the population in the three strata, we will estimate the proportion using contraception and the variance of the estimate.

The estimate of users in stratum 1 is

$$\bar{y}_1 = \frac{74}{210} = 0.35238$$

The estimated variance (from equation 3.10a) is

$$v(\bar{y}_1) = \frac{1}{7 \times 6 \times 900}\left[(11^2 + 10^2 + \cdots + 14^2) - \frac{74^2}{7}\right] = 0.00047$$

Similarly, sample mean and variance can be computed for the other two strata as given below

| Stratum | $n_i/n$ | Mean | Variance |
|---|---|---|---|
| 1 | $\frac{210}{720} = 0.29167$ | 0.35238 | 0.00047 |
| 2 | $\frac{240}{720} = 0.33333$ | 0.72917 | 0.00045 |
| 3 | $\frac{270}{720} = 0.37500$ | 0.56296 | 0.00026 |

The pooled mean (since allocation is proportional) is

```
***Example 3.8 (i):
/*We take the dataset we used for example 3.2 and add a strata variable defining
the stratification scheme given above*/

gen str = .
replace str = 1 if PSU >=1 & PSU <= 7
replace str = 2 if PSU >=8 & PSU <=15
replace str = 3 if PSU >=16 & PSU <=24

lab var str "Strata number"

**We then calculate mean and variance of contrause variable under two-stage
stratified random sampling

svyset, clear /*clearing any survey specification from before*/
svyset PSU, strata(str)|| _n /*sample selected at 2 stages, first PSU and then
with PSU randomly*/
                                /*note that stratification is defined at the first
stage*/
svy: mean contrause /*calculate mean and variance*/

[OUTPUT]
Survey: Mean estimation

Number of strata =        3          Number of obs     =      720
Number of PSUs   =       24          Population size   =      720
                                     Design df         =       21

--------------------------------------------------------------
             |              Linearized
             |       Mean    Std. Err.     [95% Conf. Interval]
-------------+------------------------------------------------
   contrause |   .5569444    .0112671      .5335132    .5803756

Note: There is no change in mean, but the variance is reduced considerably
compared to what we got in non-stratified design (example 3.2)
```

$$\bar{y} = 0.29167 \times 0.35238 + 0.33333 \times 0.72917 + 0.37500 \times 0.56296 = 0.55694$$

The pooled variance is

$$v(\bar{y}) = 0.29167^2 \cdot 0.00047 + 0.33333^2 \cdot 0.00045 + 0.37500^2 \cdot 0.00026 = 0.000126$$

$$\text{SE}(\bar{y}) = 0.01125$$

**(ii)** *Two clusters are selected from each stratum*

Suppose, two PSUs are selected from each stratum and the overall design is not an EPSEM design. By substituting the value of $a_l$ equal to 2, the sample means $\bar{y}_l$ and $\bar{y}$ can be computed using equations 3.34 and 3.35 respectively. The variance can be computed as:

$$v(\bar{y}) = \sum_{l=1}^{L} W_l^2 (1 - f_l) \frac{(y_{l1} - y_{l2})^2}{n_l^2} \tag{3.38}$$

If it is a proportionate allocation, the variance reduces to

$$v(\bar{y}) = \frac{(1-f)}{n^2}\sum_{l=1}^{L}(y_{l1} - y_{l2})^2 \tag{3.39}$$

**Example 3.9**

Consider example 3.2 and derive the estimates assuming that the 24 PSUs were selected systematically from an ordered list (i.e., as if two PSUs are selected in each of the 12 strata). The estimate of sample mean will be

$$\bar{y} = \frac{1}{720} \times \left[(11+10) + (11+9) + \cdots + (15+18)\right] = 0.55694$$

$$v(\bar{y}) = \frac{1}{720^2}(11-10)^2 + (11-9)^2 + \cdots + (15-18)^2 = 0.00035$$

$$\text{SE}(\bar{y}) = 0.01858$$

## 3.5 SELECTION OF SAMPLING UNITS AT DIFFERENT STAGES

### 3.5.1 Selection of PSUs

Apart from PPS with or without replacement, there are two ways to select sampling units by SRS at first stage, the PSUs. To suit the computation of variance estimation, PSUs are often selected with replacement. In this case, a particular PSU may get selected more than once. If number of second-stage units to be selected is $b$, then from a PSU that is selected twice, $2b$ second-stage units would be selected. The two selections, from the PSU would be considered as two independent clusters each of size $b$.

The PSUs can also be selected by SRS without replacement (Rao et al., 1962). This avoids selection of a PSU more than once. The PSUs can be ordered according to a variable that is related to the study outcome and then required number of PSUs can also be selected systematically with the help of a single random number (discussed in Chapter 2). Use of systematic sampling to select first-stage units is sometimes avoided with the apprehension that variance computation would become difficult in this case. However, it may be mentioned that variance computation, if PSUs are selected using systematic sampling, can be approximated assuming it to be a simple random sample. If PSUs are ordered by some characteristics before selecting them systematically, that is, if they are ordered with implicit stratification, the analysis can be done assuming that two adjacent PSUs are selected from the same stratum. In this case, equation 3.33 can be used for the variance estimation (Kish, 1995, p. 119; Hansen et al., 1953), even though it is an approximation. However, it is shown in the literature that the approximation would be negligible, particularly if number of PSUs in a population happens to be large.

It is prudent to examine the size of PSUs in a frame before their selection. Some PSUs may be too large and some too small. Such PSUs might need some adjustments. Let us discuss this in the context of selection of PSUs with PPS. Suppose PSUs are villages and the total population in all the PSUs is $H$ and as usual, '$a$' and '$b$' are number of PSUs to be selected and number of units from each selected PSU, respectively. The selection interval, in this case, is

$$I = \frac{H}{a}$$

PSUs of size greater than $I$ are bound to get selected, they are known as self-selecting PSUs. If there are quite a few such large PSUs, it would be better to keep them in a separate stratum. One might decide to allocate the number of PSUs to be selected proportionately to the strata and as a result keep the overall design self-weighting. Alternatively, if they are found to have some special characteristics, one can also decide to have relatively more samples from it but the selection of units within it can be done with equal probability as have been indicated earlier. The overall design will not be self-weighting but since units have been selected within a stratum with equal probability, estimates indicated above can be computed and the overall estimate can be computed by pooling together strata-specific estimates with suitable weights.

Since $b$ units are to be selected in a PSU, any PSU of size less than $b$, if selected can create a problem. If $b$ is 30 and size of the selected PSU is 15, then even if all the 15 units are selected, it will tend to reduce the sample size. If such cases are very few one can ignore them, otherwise, it is advisable to make some adjustments. In order to select $b$ units from a PSU, the minimum size of a PSU should be ideally greater than $b$. It has been discussed in the literature that the growth of villages often depends on their size; bigger villages tend to grow bigger and faster while small villages tend to shrink in size over time (Singh et al., 2008). Therefore, the minimum size of PSUs in a frame is kept as $k \times b$ (where $k = 1.5$ or 2).

There are few different ways to handle small units in the sampling frame (Kish, 1995, p. 316). One simple way to handle this is to scan the sampling frame before selection and merge the small PSUs with nearby PSUs so that total size of joined PSUs is at least $k \times b$ (elaborated in chapters 5 and 10). This procedure may bring some additional gain to the sample as by merging two or more PSUs the heterogeneity of the resultant PSU becomes higher than the individual PSUs.

However, if there are many small size PSUs and the total size of all such PSUs together is moderately large, it is advisable to keep all these PSUs in a separate strata. The number of units to be selected from a selected PSU in this stratum can be reduced suitably keeping the probability of their selection within the strata to be equal.

### 3.5.2 Selection of Second-Stage Units

Selection of sampling units at any other stage can be done with or without replacement. Selection of units, in a selected PSU, if it happens to be the last stage of selection, is generally done without replacement. The sampling technique usually used for the selection of end units is the systematic sampling as it provides greater opportunity to introduce heterogeneity in the selected sample. For example, in household listing, care is taken to number contiguous households serially. When households are selected from such a list systematically with an interval, they will be spread out and would cover the whole list. In a three-stage design, the sampling unit to be selected at the second stage is also a cluster of lower size than that of the PSU. For example, in a household survey in an urban area, wards in a town may be the PSU and next stage of selection is likely to be the CEB, particularly if the ward size is too large to handle (since selection of units at the last stage generally will require houselisting). If wards are selected with PPS and the second-stage units (enumeration blocks) also vary by size, it is desirable to select the second stage units also by PPS. Selection of equal number of households at the last stage facilitates having a self-weighting design (also discussed in Section 3.5.3). Alternatively, one can adopt equal probability selection at all stages, if variation in cluster size can be kept within the tolerable limit as mentioned earlier.

### 3.5.3 Selection of Individuals within a Household

In household surveys, selection of equal number of households at the last stage, as discussed earlier results in a self-weighting design. Often a requirement, in such surveys, is to select a member of a selected household. For a survey on fertility and family planning behaviour or on issues regarding maternal care, generally a woman in reproductive age is selected. This may be also true for many other issues like disease prevalence, sexual behaviour, gender norms, domestic violence, etc. Men can also be studied for many of these issues. The number of eligible (for interview) individuals in a household varies. Thus, if households are selected with equal probability, selection of an eligible individual can make the design deviate from being self-weighting. Three alternatives can be adopted to remedy the situation. The first is to ignore the departure from equal probability of selection and compensate the resultant bias by using appropriate weights. The second is to include all eligible members found in a selected household in the sample, to keep the design self-weighting. The last option is to select one eligible individual in such a manner that the overall probability of selection remains equal. All the three methods have some advantages and disadvantages.

The first alternative does not require any change, except that weights are to be used. For example, if '*a*' PSUs are selected with PPS using $H_i$ as the size of *i*th PSU and let $H_{ij}$ denote the number of eligible members in *j*th selected household of *i*th PSU. If a fixed number of *b* households are selected in a selected PSU and *H* denotes the total number of households $\left(=\sum_i H_i\right)$, then the probability of selecting one eligible member from *j*th household of *i*th PSU is

$$= \frac{aH_i}{H} \times \frac{b}{H_i} \times \frac{1}{H_{ij}} = \frac{ab}{H} \times \frac{1}{H_{ij}}. \tag{3.40}$$

Hence, the probability of selecting an individual in *i*th PSU will vary from one household to the other depending on the number of eligible members available in a selected household. If variation in the household size is expected to be less, that is, the departure from equal probability of selection is likely to be marginal, no change may be required. The bias can be taken care of through weighting. Selection of one individual with equal chance, though theoretically simple, is difficult to achieve in practice.

The second alternative of avoiding any selection at the last stage, that is to include all eligible individuals in a selected household in the sample, helps maintain the design to be self-weighting. However, it can make the number of units selected in a PSU to vary and hence might affect $c(x)$. Another important consequence of this decision is the intra-household correlations in a study variable. For example, many fertility- and maternal care-related variables are likely to be similar among women in a household (Ram and Roy, 2004). Unless the number of eligible persons per household is small, including all eligible individuals may not be suitable, particularly if intra-household correlation is a concern. For example, in NFHS in India, where eligible women age 15–49 years are being studied, all such women in a selected household are included in the sample. This is because the average number of such women per household is very close to unity.

The third option calls for changing the selection procedure. In this case, selection at first stage is done by usual PPS, but households, in a selected PSU at second stage, are selected again by PPS, instead of giving them equal chance. The third or the last stage selection of one eligible individual in a selected household is done with equal probability. For example, the selection procedure mentioned in equation 3.40 will change as

$$= \frac{aH_i}{H} \cdot \frac{bH_{ij}}{H_i} \cdot \frac{1}{H_{ij}} = \frac{ab}{H}$$

It may be noted that the probability of selection of a household ($H_{ij}$) in $i$th PSU is not independent of its size as was apparent in equation 3.40. Selection probability of $j$th household in the $i$th PSU is also in proportion to its size (PPS). In this case, the design remains self-weighting as well as only one individual is selected in a selected household. The only disadvantage, in this alternative, is that it requires knowledge about size of households in a selected PSU before its selection. In other words, household listing operation will entail, apart from enumerating all households with their addresses, recording number of eligible household members. For example, if the eligibility criteria require selection of adults, at the time of listing, number of adult members of a household will need to be obtained during listing. This may increase the cost of household listing operation. It may be mentioned that a listing operation receives less importance than actual fieldwork. This is natural, because the listing operation normally entail only counting number of households and record them in a specified geographic order. A field investigation, where all necessary information is collected, requires proper rapport building and taking appropriate consent including providing a broad idea about the study in a selected household. No such rigorous formality is necessary in a household listing operation. Collecting some information about a household, even if apparently not sensitive, during listing is liable to have errors, unless the household listing staff are trained suitably, increasing the cost of training of field-staff for household listing (also discussed in chapter 4).

In both alternatives 1 and 3, it is required to select one eligible individual giving equal chance to each in a household. The process of randomness can easily get violated if not implemented properly. Inclusion of an individual, selected with equal chance, depends on availability and willingness of the individual. If a randomly selected individual is not available or unwilling to participate, whereas another individual, not selected, is both available as well as agreeing to take part, there will be a temptation to consider the latter. First, it is difficult to visualize the consequences of such departures and hence one gets inclined to take the easier route of deviating from the strict randomness. Also, often the quality of work of an investigator depends on how efficient she/he is in restricting non-response. Unless there is proper realization of the importance of randomness and adequate supervision of selection procedure with simple and clear rules for selection is followed, it may be difficult to implement strict randomness in selection of one individual. Giving instructions such as to select a random number and hence select one individual is inadequate. Providing a set of random numbers to each interviewer that should be used in a strict order is helpful but not adequate as the interviewer can easily change the order of individuals in the sample. Therefore, apart from providing a list of random numbers, interviewers should be instructed to order individuals in a specific manner. Such as start with the oldest member who should be given serial number 1, then the next oldest number 2 and so on. Selection in this way can work and good supervision can detect the departures from randomness.

Inability to randomize can create problems that can be serious in some situations. For example, in a survey on prevalence of a particular disease, if one individual from each selected household is chosen not with strict randomness, it might end up having more persons suffering from the disease into the sample because they are more likely to be present in the household. This would overestimate the prevalence. In a survey on domestic violence against women, it might be true that those who have experience of or are currently experiencing such violence would be less willing to participate due to fear of repercussions. It can be the other way also, that is the affected women may be more forthcoming to report than who have not experienced violence, may be with the hope of getting some solution. Whichever way the bias in selection happens, it would tend to create problems during analysis and will be difficult to take care of. At the least, such departures from randomness can make the sample less representative of a population. For example, in a tobacco use survey, where eligibility of the survey is all adults, may end up having unduly high proportion of older people (age 65 and older) because they are more likely to be present in the household. Since tobacco use declines at older ages, estimated prevalence is likely to be underestimated. This type of bias can be adjusted by post-stratification (discussed in Chapter 2).

Kish (1949) has provided tables, known as Kish tables, which are used widely to select one individual from a household. One small table is attached to each individual questionnaire. The method requires that each eligible member of a household is listed in a particular order. Once this ordering is done and the number of eligible person in a household is known, the attached table can be referred which fixes the serial number of the individual to be selected. Implementation of Kish table is simple and it attaches equal probability to each eligible member for inclusion in a sample. Since total number of eligible persons ($H_{ij}$) is known, equation 3.40 can be used to obtain the overall probability of selection of a respondent. Computer Aided Personal Interview (CAPI) technique, used as an alternative to paper-pencil personal interview can automatically select an eligible respondent by generating a random number.

## 3.6 KEY POINTS

The chapter presents sampling designs where selection of units is done in more than one stages. It is by far the most widely used design, particularly for household-based surveys. Some key points are

- In a multistage design, a sampling unit that is selected at first stage, known as the PSU occupies a pivotal role.
- Estimation of a parameter is done at PSU level. A selected PSU is often referred to as a cluster. The values of a characteristic $y_i$ are cumulated for each cluster, irrespective of the stages of selection. Similarly, the count variable $x_i$ is also obtained for a cluster by summing the number of units selected in it.
- Estimate of mean or proportion is usually a ratio. The ratio and its sampling variance can be obtained as shown in equations 3.14 and 3.15.
- While the ratio obtained from a sample is not an unbiased estimate of the population value, it is generally a good approximation. The estimate of sampling variance is also an approximation and its validity depends on the coefficient of variation of $x$ ($c(x)$), that is how $x_i$, for a given mean number of units per cluster, varies over different PSUs (equation 3.19).
- It is desirable that a design yields a value of $c(x)$ less than 0.1 However, a value of 0.2 is also tolerable.
- Units within a cluster tend to be related. The presence of intra-cluster correlation influences and reduces the number of samples that are identically and independently drawn.
- The presence of intra-cluster correlation tends to increase the sampling variance of a cluster design compared to a simple random sample of same size. The design effect (deff), which is a ratio of the two sampling variances, is given by

$$\text{deff} = 1+(b-1)\rho$$

- where $\rho$ is the intra-class correlation coefficient.
- While planning a study, an appropriate choice of PSUs is required. It is better to keep the size of PSUs as small as possible, which means the number of PSUs should be as large as possible in a population.
- In sample selection, it is beneficial to have as many PSUs in the sample as possible, keeping the resource constraint in mind. This can help in reducing deff and the sampling variance and also facilitate controlling the value of $c(x)$.
- While designing, care is needed if PSU sizes vary considerably in a population. The large variation in cluster size (number of units selected in a PSU) is detrimental for the design.
- The possible remedies to control the cluster sizes are i) to adjust PSU sizes suitably before selection so that their variation is reduced, ii) stratify the PSUs by size, so that within a strata variations in PSU size reduces, and iii) select PSUs by PPS or PPES (with estimated size of PSU if their sizes are not available) and within a selected PSU, selection of similar number of units, with EPSEM.

Chapter 4

# Probability Sampling under Imperfect Frame

## 4.1 INTRODUCTION

Availability of a sampling frame is a basic requirement for application of a probability sampling technique. In order that each and every element in a population has a known and non-zero chance of being included in a sample, an ideal frame should consist of all elements occurring only once and it should exclude any other element that is irrelevant for the study. The basic techniques, such as SRS, stratified sampling and systematic sampling, discussed in Chapter 2, presume availability of such a list of elements if elements are being selected directly. In cluster and multistage sampling, a list of all areal units comprising a population is required initially. They are often selected using a PPS sampling (also discussed in Chapter 2), which assumes an availability of size or at least an estimated size of each cluster. For selection of elements in the final stage of a multistage design, a frame consisting of elements in selected areas would be required. This is generally obtained through listing of dwellings or households in a population-based survey.

It is indeed difficult to have a perfect frame in every situation. Let us consider, for example, a common situation of studying a population having certain characteristics, say, suffering from a particular disease. A usual available list will consist of both the elements, households having a member suffering from the disease and those without anyone suffering. The latter element may be irrelevant and, therefore, considered as a blank.

The present chapter discusses a few applications of probability sampling in the absence of an ideal frame, following three broad situations that are generally encountered:

**(a)** Sampling populations having specific attributes, that is, a frame consisting of elements as well as blanks.

**(b)** Sampling populations using a defective frame, that is, a frame which is either incomplete or consists of duplications (same element occurring more than once).

**(c)** Sampling populations in the absence of any frame.

In addition, the chapter also includes a discussion on household listing that is advisable in implementing a cluster sampling.

## 4.2 SAMPLING POPULATIONS HAVING SPECIFIC ATTRIBUTES

This situation arises when a survey objective refers to not all the elements in a population but to a section of it. For example, the objective may be to study elderly or for that matter any other age or sex group of a population, migrant population, population suffering from certain diseases, persons belonging to specific socio-economic or ethnic groups, etc. (McKenzie and Mistiaen, 2009). Elements in a population can be divided into two groups. One group consists of elements possessing the attribute under study and hence regarded as a part of the target population. The other, not having elements with the attribute, is considered as blanks.

Presence of blanks, if they are not screened and excluded before selection, can cause a few problems. Their selection in a sample may reduce the effective sample size. It will also tend to increase the sampling variance of an estimate. On the other hand, screening requires additional resources.

Sampling techniques to be adopted in such cases often depends on the proportion of persons with the attribute in a population. It can be noted that smaller the proportion of individuals with the attribute, larger will be the size of the blanks. For example, if migrant households consist of 1 per cent of a population, one needs to screen an average of 100 households to get one migrant household. If proportion of population with an attribute is less than 1 per cent, it is generally regarded as a rare attribute (Kish, 1995, p. 404–405). Because of the cost implications, study of populations with rare traits is often a problem.

At this stage, it is pertinent to distinguish between the study of a population with a rare trait and the estimation for rare or small subpopulations. The two can be different. The former, as mentioned, refers to a subsection of a population with an attribute. When that subsection is small, smaller than 1 per cent, we tend to call it a subpopulation with the rare attribute.

The second issue is an estimation of an attribute (not necessarily rare) for a smaller area or subpopulation. For example, in a national survey, it is often a concern to provide estimates, in addition to the national estimate, for its different administrative subdivisions (e.g., districts, blocks, etc.). If the national sample is drawn with an EPSEM, The number of elements selected from a block may not be sufficient to provide a valid estimate for it. This is particularly true if proportion of the national population living in the block is small. This is referred to as estimation of rare subpopulations or small area estimation.

Our aim here is to discuss the first aspect, that is, survey-based estimation of characteristics of a population having a specific attribute. Two different objectives can be considered. One is the estimation of the proportion, that is, proportion of a population possessing the attribute. For its estimation, we need to sample both the groups, persons possessing the attribute as well as those not possessing it. The second is about studying certain characteristics of the subpopulation possessing the attribute. Persons without the attribute are clearly irrelevant in this case and hence to be regarded as blanks. The procedures for estimation of rare events are discussed subsequently.

### 4.2.1 Sampling when Target Population is not Rare

Let the total population be $N$ in which $M$ belongs to the target population, possessing a particular attribute. The determination of sample size will depend on the specific objective of the study. Let us assume that it is both to estimate the proportion belonging to the target population and to study a specific characteristic of the target population. For example, the objective may be to estimate the proportion of persons in a population suffering from a disease and also their treatment-seeking behaviour. For the latter objective, the target population is those suffering from the disease.

There are two distinct approaches that can be adopted, particularly for studying the latter objective.

#### 4.2.1.1 Sampling without screening

The desired sample size $n$ can be estimated as discussed in Section 2.2.7. While deciding on the sample size $n$, one can ensure that $n$ will provide an adequate number of persons, say, $m$, who belongs to the target population. If an EPSEM sample of size $n$ is selected, one expects that $m$ will be equal to:

$$m = n\left(\frac{M}{N}\right)$$

The usual multistage design discussed in Chapter 3 can be employed to select a sample of size $n$ with a household listing at the last stage to select units. The estimation of the proportion of persons suffering from the disease ($M/N$) and treatment-seeking behaviour of those suffering can be done based on the sample.

Let,

$x_i$ = size (number of persons) of $i$th PSU

$y_{ij}$ = 1 if $j$th person in $i$th PSU is found to be suffering from the disease (possessing the attribute)

= 0 if $j$th person is not suffering from it (does not possess the attribute)

$z_{ij}$ = 1 if $j$th person in $i$th PSU suffering from the disease and sought treatment (the characteristic under study)

= 0 if $j$th person in $i$th PSU suffering from the disease and did not seek treatment

$y_i = \sum_{j=1}^{x_i} y_{ij}$ = Number of persons suffering from the disease in $i$th PSU

$z_i = \sum_{j=1}^{x_i} z_{ij}$ = Number of the persons suffering from the disease who sought treatment.

The estimate of the proportion of persons suffering from the disease ($r$) will be a ratio estimate:

$$r = \frac{\sum_{i=1}^{a} y_i}{\sum_{i=1}^{a} x_i} = \frac{y}{x}$$

where $y$ denotes the number of persons suffering from the disease ($m$), $x$ is the total number of persons selected ($n$) and a is the number of PSUs selected.

The estimate of the proportion (among those suffering from the disease) who sought treatment ($r'$) is a subclass estimate and it can be estimated as a ratio

$$r' = \frac{\sum_{i=1}^{a} z_i}{\sum_{i=1}^{a} y_i} = \frac{z}{y}$$

where $z$ denotes the number of persons suffering from the disease who sought treatment. The sampling variance of $r$ and $r'$ can be obtained as mentioned in Section 3.3.1:

$$v(r) = \frac{1}{x^2}\left[v(y) + r^2 v(x) - 2r\,\mathrm{cov}(yx)\right]$$

and

$$v(r') = \frac{1}{y^2}\left[v(z) + r'^2 v(y) - 2r'\,\mathrm{cov}(zy)\right]$$

If stratification is used before selection of PSUs, equation 3.28 can be used to estimate variance of the ratio mean.

#### 4.2.1.2 Sampling with screening

If the objective is only to study a characteristic of the study population and not their proportion, a slightly different approach can be taken. In this case also, a multistage design (Chapter 3) can be employed. Assuming that a two-stage design is adopted, selection of PSUs can be done as usual, preferably by using probability proportional to their size. Screening of households to identify blanks is to be undertaken at the time of household listing for each selected PSU. That is, in addition to recording addresses of each household in a selected PSU, the household listing staff also record the necessary information to categorize each household as eligible (i.e., possessing the attribute being studied) or not eligible, that is, a blank. Selection of households at the last stage may then be carried out by considering only the eligible households identified by the listing exercise. If the sample is selected in more than two stages, the same principle may be extended. The household listing is done in an areal unit that is selected before selection of households (elements). The procedure of estimation of mean (or proportion) of a characteristic of the target population (a ratio estimate) and its variance can be done as mentioned in the previous section.

It may be mentioned that, though it is not specifically planned, if required, an estimate of the proportion of persons having the attribute can also be done by using the PSU level information on number of eligible and blank households carried out at the time of household listing.

#### 4.2.1.3 Relative advantages with screening and without screening

One constraint in screening of households at the time of listing is that it tends to increase the cost of listing. Household listing as such requires enumerating all structures and households in a PSU in a specified order. The procedure can be completed fairly quickly as not much information is generally collected. However, for the purpose of a screening if some in-depth information on one or more attributes are to be collected from a household on their members, listing staff will need to have interviewing skills and thereby increase the cost, both in terms of time and staff compensation. We explain these situations with a couple of examples.

Suppose in a study the target population is poor people staying in *Kachcha* houses (e.g., houses made of mud) or in another, households having an elderly member (more than 65 years of age). In the first case, a mere observation can help in identifying the type of house and hence can facilitate separating blanks from the eligible population. In the second case, identification of an elderly member in a household may just need a simple question (whether a person of age above 65 lives in a household) and does not complicate the listing procedure. In comparison, consider an example where the objective of the study is to select persons suffering from a stigmatized disease or persons having a criminal case pending against him. In such cases, a person who is proficient in interviewing skills would be required. Further, ascertaining target population in such cases would require sufficient time to be spent in a household to build rapport and explain the purpose. Listing operation, which is done in a fairly large scale (it is generally proposed that around 20 units should be listed in order to sample one unit, Kish, 1995), would thus become inefficient. Even an issue, like identification of a migrant household, can be a sensitive one in a specific situation which may adversely affect the process of screening at the time of household listing. A choice between the two alternatives depends on proportion of persons with a specific attribute and type of screening that would be required to identify persons with the attribute.

The screening method provides better control over selection of number of persons with an attribute ($m$). This is particularly helpful if $P$ is small. For example, suppose $P = 0.2$ and one needs to have a sample of 250 eligible persons. If screening is to be avoided, one needs to draw at least 1250 households and still may not be sure to obtain 250 cases. However, if screening is done before the selection of cases, a required sample size can be ensured. Hence, if screening is not elaborate or does not deal with any sensitive issues, it is generally preferred.

#### 4.2.1.4 Facilitating screening

It needs to be emphasised that errors made in screening of elements can introduce errors in estimation of a parameter. Two types of error are possible while screening. An element, which is actually a blank, is screened as an eligible unit or the other way round, that is, an eligible unit is recorded as a blank at the time of screening. Chance of occurrence of such error increases if the attribute under study is sensitive or deals with certain phenomenon that carries stigma. In such cases, it is natural to expect that error of the second type will generally occur, particularly if adequate care is not taken at the time of listing. Error can also occur due to ignorance. For example, in rural India, people generally do not keep track of their age and it is often difficult to ascertain their correct age. In a survey on elderly (persons of age 65 and above), if a listing is done, there can be an error and it can be of both the types. That is, an ineligible person (who is actually below age 65) can be screened and recorded as eligible (an elderly) and vice versa.

The question that is pertinent in sampling is the relative role of the two errors. If an ineligible person is recorded as eligible, it might get selected and, if so, will become a blank and reduce the sample size. In other words, it will tend to increase the sampling variance of the estimate. Error, in the other way, means an eligible person will not get a chance (because he is recorded as a blank) of being included in the sample. This can create a bias, and, as has been mentioned, bias is always more problematic unless its magnitude is known. Although effort should be made to control both, extra care, if possible, should be taken to avoid bias. In the above example, it would be better to keep the age limit as 60 years and above instead of the required 65 and above. This will increase the chances of having some blanks into the sample but will reduce the chances of an eligible person getting recorded as ineligible. The reduction in sample size because of this can be compensated by suitably increasing the sample size in advance.

There are ways to reduce the screening cost. For example, the need for revisits to households that were not available can be avoided. If the characteristic being screened is not sensitive and is likely to be known to neighbours the information can be sought from neighbouring household (Kalton, 2009). In order to minimize the possibility of incomplete coverage, if a neighbour informs absence of such a person (in the household which happens to be not available), the same enquiry is again made from the next household also. Note that no such confirmation is made if the neighbour's response is 'yes', that is, someone with the characteristic lives there. As mentioned above, even if there is a discrepancy in the neighbour's reporting, it will tend to influence the error and not the bias.

This concept of acquiring information about a household from the neighbour can be extended to reduce the screening cost. For example, if $L$ is the total number of households in a PSU being screened, the actual screening can take place in $L/3$ households with each screened household providing necessary information for two neighbouring households on either side. To keep track of under coverage, a comparison of proportion of households with the attribute in actually screened household and those that are obtained indirectly can be compared (Kalton, 2009). For example, suppose a PSU has 198 households. If the listing is done correctly the proportion of households with the attribute in 66 households actually screened and in 132 where it was reported by a neighbour are likely to be similar.

## 4.2.2 Sampling for Rare Attributes

When a characteristic or event being studied refers to a small proportion (approximately <1 per cent) of a population, it is referred to as a rare population. As mentioned, the problem here is that one needs to have a large population or screen a large number of households in order to have one eligible household, a household having or possessing the rare trait. This tends to render the household-based sample selection, which happens to be a very desirable approach for application of probability sampling, inefficient. We first discuss the situations where it would still be feasible to employ a household-based sampling procedure.

### 4.2.2.1 Household-based sampling of rare population

A household-based sample selection can be adopted in following situations:

1. **When screening is quick and simple**: If quick screening is feasible, one can still consider a large population in order to get the required sample of cases. For example,

to get 500 cases of a rare population, if its proportion is 0.5 per cent, one needs to screen about 100,000 households. It is advisable to consider PSUs of larger size than is usually preferred. Larger cluster size is generally avoided to reduce the effect of clustering or the design effect (Section 3.2.2.3); however, in this case, even if total number of households in a cluster is, say, 1000, the actual size, number of them having someone with the rare trait will be small. It will be, on an average, five households per cluster. Screening 100 PSUs each with 1000 households can give the required sample size. Whenever feasible, stratification of PSUs, at least, by prevalence of the attribute, if known or variables related to it would be desirable.

2. **When there is geographic clustering of rare population**: Geographic clustering or the tendency to stay together is quite common, particularly in the case of rare ethnic groups, high income individuals, etc. For example, to estimate persons, in a city, in 'high income' category, which may be of rare occurrence, there are areas where such households are generally located. For certain rare diseases, there are locations which are more prone to the disease than other areas. If such geographic clustering is present, sampling based on household selection is facilitated by stratifying the geographic areas according to the prevalence of the rare population. The sample can be allocated disproportionately with greater coverage from areas having greater prevalence. This approach can lead to greater efficiency. It has been shown (Kalton, 2001) that gain in such disproportionate allocation depends on:

   **(a)** relative proportion of a rare population in a stratum compared to that in the total population, and
   **(b)** degree of concentration of the rare population in the stratum, that is, what proportion of its total population can be found in that stratum.

   Let us elaborate this with an example. The data given in Table 4.1 show that the distributions of population in its two strata by composition of persons with a rare attribute differ substantially. Not only that the population with the rare attribute is much higher (0.02 per cent) in stratum 1 but it also constitute of more than 90 per cent of persons with the attribute in the population. Larger gains can be obtained by disproportionately allocating larger sample in stratum 1 in such a case.

TABLE 4.1 Distribution of total population, population with a rare attribute and per cent of population with the attribute in a population divided into two strata.

| Strata | Population | % with rare attribute | Number of persons with the attribute |
|---|---|---|---|
| 1 | 50,000 | 0.02 | 1000 |
| 2 | 200,000 | 0.0005 | 100 |
| Total | 250,000 | 0.0044 | 1100 |

However, the gain also depends on the cost of data collection from a household with a rare population relative to that of screening a household not having a member of the rare population. The gain tends to decline as cost of collecting data from rare population increases in relation to that of the screening (Kalton, 2001).

When there is geographical clustering, like that depicted in Table 4.1, it is possible to consider slight modification of the objective. Instead of studying persons with the rare attribute in the total population, it can be undertaken only for the population in stratum 1. One can, considering cost implications, avoid stratum 2 totally. Even if modification of the study objective is unsuitable, screening for stratum 2 can be aptly changed. An idea about whether a member of a rare population resides in a selected cluster in sparsely distributed area can be obtained by selecting randomly few or even one sample household from it (Waksberg, 1978). If yes, detail screening of the cluster can be undertaken. This requires the assumption that everyone within a cluster is likely to know about the particular attribute and presence of a person with the attribute within the cluster. One can modify the screening procedure and decide about whether to screen a selected cluster by asking a few responsible persons in it. For example, in the context of rural India, one can consider a village (which are generally considered as clusters) panchayat member (responsible for administrative activities of a village), a school teacher, a doctor, etc. to determine whether a selected village will have any member of the rare population and hence should be screened or not.

3. **Two-phase sampling**: In a two-phase sampling, information is collected at each phase. Information collected from sampling units selected at first phase facilitates selection of units at the second phase. This approach can be fruitfully employed irrespective of whether there is geographical clustering of a rare population or not. This is particularly helpful when screening involves an elaborate procedure.

   An example may help to clarify the procedure. The objective of a study was to estimate the prevalence of gallbladder diseases and the risk factors associated with it in a Gangetic belt in the North-Eastern region of India. It is a rare disease and it requires expensive screening with the help of an ultrasound machine carried out by a trained paramedic or preferably a doctor. Hence, large-scale screening for this rare disease was ruled out. The selection of PSUs (villages), at the first phase, was done as usual after stratifying them considering a number of factors related to the disease, including village size. A questionnaire consisting of attributes that are related and can be considered as symptomatic for the disease (such as whether a person has suffered from jaundice, have had a stomach ache of a particular type in the recent period, etc.) was canvassed in all households of a selected PSU. The questionnaire also elicited information about other characteristics that are relevant for understanding the risk factors. Based on the information collected at the first phase, households in a selected PSU were stratified into two groups as 'symptomatic' and 'asymptomatic'. Since symptomatic cases were expected to have an elevated risk of the disease, all such cases, which was generally small in number in a PSU, were included for the ultrasound testing at the second phase. Asymptomatic cases were randomly selected in a 2:1 proportion for the testing also at the second stage. That is, for each symptomatic case there were two asymptomatic cases. The selected sample was weighted accordingly for estimating the desired characteristic. The procedure facilitated achieving both the objectives of the study (Unisa et al., 2011).

4. **Use of available list**: A household-based probability sampling of rare population becomes a problem when screening is not simple and the attribute happens to be more evenly distributed. If a frame or a list of households with a rare attribute is available the usual approach can be easily adopted. For example, one may be interested in a survey of practising doctors in a city or select a sample of car owners. In both the cases it might be possible to use a probability sampling (discussed in Chapter 2) with the help of the available registration data. However, one needs to check the list for possible errors before use. In the first case, it may be possible that the list may contain number of blanks in the sense that some of the doctors may no more be staying in the city or may have completely stopped practising. This will be a situation where a frame consists of blanks and the procedures to deal with blanks are discussed above. In some cases, particularly for the latter, a person can have more than one car. In other words, the list may have duplications. The possible alternatives in such cases are discussed subsequently. A major problem can arise if a list is incomplete. As already mentioned, this can be a more serious error than the other types of error because it has potentials to introduce bias. Bias will be introduced if those that are left out of the list form a substantial proportion of the population.

   Apart from the magnitude of incompleteness, another important issue is how similar is the group that is being left out from those that are in the list. Let us consider an example of estimating average tax paid by households among those that pay taxes in a city. Suppose the frame is not up-to-date and consists of missing tax-paying households. If the average tax paid by the missing households happens to be similar to those that are in the list, the estimate from the list will not be affected much. But if the missing tax paying households come systematically from lower income group (who pay less taxes), the sample estimation of average tax paid is likely to be overestimated. However, it may be noted that the estimation of total taxes paid by the population may not be affected much. The total amount of taxes paid by the lower income group is not likely to be substantial, unless of course a large proportion of them are missed. However, if omission is from the upper income bracket, estimates of both mean and total income will be underestimated. Hence, if the missed cases are different (not a random sample) from the available cases, mean is likely to be biased. The magnitude of bias will depend on the extent of the departure and size of the missing units. The bias in the estimate of the total will depend, apart from the magnitude of missing units, on the characteristics of the missed values. If their values are small, the extent of bias will be less. The methods to handle incompleteness through supplementary listings have been discussed subsequently. Situations also exist where frame is not available at all. A time-location survey, discussed later, can be employed if members of a rare population are likely to be found in certain locations or at some special occasions.

## 4.3 DEFECTIVE FRAME

As mentioned, the term 'defective' refers to duplications and omissions in a frame. Another type of flaw can exist in a frame consisting of foreign elements or blanks. Methods to deal with blanks have already been discussed.

### 4.3.1 Duplications

In the following we discuss how duplication of elements in a list can influence estimation of a parameter and different approaches to tackle the problem. An element can occur more than two times in a list. This can be dealt with in the same manner as the duplications.

#### 4.3.1.1 Estimation in presence of duplications

The larger the number of appearances of an element in a list, the higher would be the chance of its inclusion in a sample. Selection of elements from such a list can create bias in estimation unless appropriate steps are taken at the time of either selection or estimation to take care of the varying chances of selection of elements. If the units having higher chances belong to a separate group or have distinct characteristics from others, its estimation would be affected. Since the group is likely to get unduly high representation, it would bias the estimation of parameters of interest.

Let us clarify this with an example. Assume that one is interested in estimating average income of persons in a population where everyone has a car and hence uses the car registration data as the list for selecting a sample of persons. It may be mentioned that use of such a list is not only convenient but also a cost-effective approach for the sample selection. In this case, if a person has three cars, her/his chance of inclusion in a sample is three times more than another person having only one car. Because of this, it is natural to expect that a sample, if selected randomly from such a list, would have higher representation of persons having more than one car. It is also natural to expect that possession of number of cars by a person is directly linked with her/his income. Hence, the selection procedure would unduly favour higher income groups to be more in number and the procedure of estimation will be biased and would overestimate the mean income.

It needs to be noted that the biasing effect of a selection approach using a list consisting of duplications depends on the characteristic under study. Since income, the characteristic under study, is directly related to the procedure of selection (possession of number of cars), the effect is likely to be large. But if the objective is not to estimate income but to study their opinion on some issues, the effect may be marginal. Particularly, if the opinion of a person is not related to the number of cars one possesses.

The issue of duplications is commonly faced while studying mobile population. The estimation where an individual is not linked to any fixed residence, such as in household-based surveys. Such cases can include: i) persons visiting a shopping centre, library, a health facility or any religious or community-based activities etc., ii) number of cars or any characteristics of persons driving a car that crosses, for example, a toll plaza, and iii) number or characteristics of nomads who congregate at some point, etc. In all such cases, an individual in a population may be found visiting a place more than once during a survey period and hence has a multiple chance of being included in a sample.

Discussion on how to facilitate employing a probability sampling in such cases, where no frame is generally available, has been incorporated in Section 4.4.

#### 4.3.1.2 Procedure to deal with duplicate listing

The reduction in cost of a survey is generally the biggest advantage in using a readymade list. Hence, even if a readily available list contains some duplications, considering the cost aspect it may be beneficial to use it with modifications either at the time of selection or at estimation. Before mentioning the modifications, two other issues need to be discussed. In some situations, the objective of a study may be such that a list containing duplications can be used as if it is without any modifications.

For example, consider estimation of an average number of lottery tickets purchased by a person by selecting a sample from a list of lottery tickets sold (assuming that a person's identity is recorded in a ticket when sold). In this case, it becomes relevant to select a sample of tickets without bothering about how many tickets a particular person has bought. If two selected tickets belong to one person, the person's frequency of purchasing the ticket will be two. In other words, the number of tickets purchased by a person will become her/his weight. One can have a frequency distribution of selected persons to get the average number of tickets a person has.

Similarly, the objective may be to select a sample of visitors to a health facility to understand different aspects of the services provided by the facility. During the period of the survey, a person may make more than one visit to the facility and hence may be included more than once in the sample. It is possible that the topic of interest is about the availability of a specific service (say doctor's presence) at the time of a visit and hence more than one visit by a person might be acceptable.

In other words, in view of a specific objective, an apparent duplication may not require any modifications. This is because, for example, in the case of lottery ticket problem, the element is not a person purchasing a lottery ticket but a 'lottery ticket' itself. Similarly, in the other case, a 'visit' is the element and not a person making that visit. It may be mentioned that in both the above cases, a household-based survey is possible. For example, one can select individuals in selected households in two stages. From selected persons, information can be collected about number of lottery tickets purchased by her/him or in the latter case about the availability of a doctor.

The other issue pertains to cleaning a frame before using it. If persons are the elements and if their duplications, in a list, need rectification, an obvious choice is that of cleaning the list. This entails removing all duplicate appearances of an element. For example, in house listing in a selected PSU, a person may have two houses within the same PSU. In India, this is not very uncommon in rural areas. Instructions are given to a listing team to list one of them (where the person spends most of his time in a day) as 'eligible' and the other should be recorded as 'ineligible' or blank. However, often it is not feasible to undertake the task of cleaning a list. Two approaches to make amends are discussed below.

1. **Correction before selection**
   To avoid duplications, it is possible to specify a rule for selection. As mentioned above, a rule has to be specified to select one household if a person has two or more within a PSU. In the case of selecting a person from car registration data (as in the example above), the specification may be made, for example, by considering time of purchase, say, the one that is purchased most recently. That is, if a person has purchased two

cars, one in 2006 and the other in 2012, the listing in respect of 2012 becomes eligible for selecting the person and the other would be treated as ineligible or a blank. In this way one can avoid giving varying chances to units for selection. In a survey on elderly, if a design requires selection of one elderly from each household, the specification may be to select the oldest among the olds, if there are more than one elderly. It needs to be noted that the resultant age distribution of elderly in a sample, in this case, would not agree with that in the population, particularly if number of households having more than one elderly is substantial in the population. Alternatively, one can use a Kish table for the selection (Kish, 1995, p. 398–99). For selection of persons visiting a library, the rule can be to select the person in a selected visit if that visit happens to be his first during a survey period. The blanks will tend to increase the sampling variance of an estimate.

2. **Correction at the time of estimation**

If duplications cannot be avoided, it can be adjusted at the time of estimation by using appropriate weights. Let us discuss this in the context of estimation of a characteristic among visitors of a facility. The visits can be selected with an EPSEM and required information on a characteristic, for example, $y_i$ for the $i$th selected visit, can be obtained. In this case, a person randomly selected in a visit will have a higher chance of selection if he has made multiple visits during the survey period. Let the probability of selection of $i$th person be $p_i$ and number of times he is likely to visit the facility is $t$. If $t$ is not known, it needs to be ascertained at the time of interview (e.g., the questionnaire may include a question on expected number of visits during the survey period). Hence, his weight $w_i$ will be

$$w_i = \frac{1}{tp_i} \tag{4.1}$$

The mean ($\bar{y}$) of $y_i$ can be obtained as

$$\bar{y} = \frac{\sum_1^n w_i y_i}{\sum_1^n w_i} \tag{4.2}$$

where $n$ is the number of persons selected. Considering $\bar{y}$ as a ratio estimate, its variance can be obtained from equation 3.15, if no stratification is used and equation 3.28 if stratification is used.

The above method requires a fairly good idea about $t$. As will be discussed later, in some situations it may be difficult to obtain $t$. The estimate will become unreliable under such situations. In the above example, let $N$ be the number visits, $M$ the number of persons making the visits in a population, and $n$ and $m$ the corresponding values in a sample. If the sample is selected with an EPSEM, it can be shown that

$$m = M \times \frac{n}{N} \tag{4.3}$$

From a sample, it can be seen that $m$ will consist of persons who visit once ($m_1$) and those who visit or intend to visit $i$ times ($m_i$). We assume that more than two visits are negligible (to ensure this, at the time of implementation, the fieldwork time is often adjusted accordingly). The value of $m_2$, that is, those with two visits consist of two types: one who appears in the sample only once (i.e., even though a person visits twice, he is selected by chance only once) and the other appearing in the sample twice. Let the first type be denoted as $m_{21}$ and the second as $m_{22}$. It has been shown (Kish, 1995, p. 392–395) that

$$\frac{m_{21}}{m_{22}} = \frac{1-f}{f} \tag{4.4}$$

where, $f$ is the sampling fraction ($n/N$). It is advisable to check that both equations 4.3 (if value of $M$ is known) and 4.4 hold. Such checks can provide an idea about the presence of bias in estimation of $t$.

#### 4.3.1.3 Incompleteness or omissions in a frame

If a sampling frame is incomplete, there is a need to take appropriate steps to avoid creating a bias in an estimate. This is true for a sampling frame used at any stage of sampling. The incompleteness can be ignored only if the magnitude of non-coverage is small, around 5 per cent, and the characteristic of the non-covered population is not very different from those who have been covered (discussed in Chapter 5). In a cluster sampling, to select households in a selected PSU, one needs a sampling frame that can provide an updated list of all households in the PSU. In many situations, a readymade list may already be available which is updated fairly recently. A simple linking procedure might then take care of the incompleteness in the list. This procedure is described subsequently while discussing household listing.

Another popular approach to deal with incompleteness is to make use of an additional list (although use of more than two lists is possible, we confine our discussion to two lists) to ensure that together they cover the entire population. For example, sampling using a dual frame was used in the National Survey of America's Families (NSAF). The primary frame for the survey was the telephone directory. The sample selected with the help of telephone directory was supplemented by the use of an area sample to cover households without telephones (Waksberg et al., 1997). It was known that about 20 per cent of families in poverty were without telephones. Since one of the objectives of the survey related to children of families in poverty, the estimates based only on telephone directory would have been biased. On the other hand, a survey based only on area sample where households would be selected irrespective of whether they have telephone or not could give unbiased estimate but at a much larger cost. As another cost-saving measure, the survey, while selecting samples from the area frame omitted areas from the frame where the proportion not having telephone was very low.

The use of an additional sample from a different frame as a supplement can be made profitably in many situations. For example, to plan a low-cost survey to obtain opinion on certain issues from professionals, such as doctors or demographers in a city who constitute, if not rare, a very low proportion of the total population. It is possible that a register would be available that could be considered as a viable primary frame. A sample drawn from such a list can then be supplemented, if necessary, with an area sample or any other list that is found suitable.

It is possible to get duplications in such cases. The same doctor listed in the register can also be selected in the area sample. Once the sample sizes to be selected from the two lists are decided, it would be better to select the area sample first and then it can be checked, if feasible, that the selected doctors do not remain in the primary list. That is, names of all those selected through the area sample can be deleted from the primary list to avoid giving them more than one chance of occurrence. It would be difficult in the other way. If the sample from the primary list has been selected first as it would be more difficult to locate and exclude them from the area sample which would be quite scattered in the area. One can consider the duplications and make necessary correction at the stage of estimation by using weights, as discussed above. The sample size for the two lists can be based on two considerations (Waksberg et al., 1997):

**(i)** proportion of the study population covered by the two sources: larger the coverage of a list, greater should be the sample size allocated to it, and

**(ii)** cost of data collection using each source: higher the cost, less should be the sample size selected from it.

Another factor that needs to be considered is the type of sampling to be adopted to draw samples from the two lists. In the case of an area sample, cluster sampling is likely and, therefore, one needs to consider the design effect. Considering the two samples to be from two different strata, the two estimates can be pooled using their weights as mentioned in Chapter 2 (Section 2.3.2.1).

One can also consider employing the dual-frame approach for the estimation of infant mortality or even maternal mortality, which requires large sample sizes, particularly for the latter. In this case, for areas that have fairly good birth registration or have high proportion of births occurring at institution, such records can form the primary source which can be supplemented by the use of area sample to take care of the incompleteness. For populations where institutional birth is 80 per cent or more, one may even consider the possibility of avoiding the use of an estimate from the supplementary sample, the area level sample. For example, if some valid research is available to establish a relationship between infant mortality or maternal mortality based on births occurred in institution and at home, it may be used to compensate for the non-coverage.

## 4.4 SAMPLING IN ABSENCE OF A FRAME

This section deals with situations where a household-based survey is ruled out mainly because of its cost implications or the difficulty in locating the units. For example, for studying characteristics for persons with HIV, males having sex with males (MSM), etc. fall under such categories. Consider those engaged in prostitution. There may not be any register or roster available for their selection. The nature of their profession is also such that it would generally be difficult to identify them in a household-based survey. The same is also true for studying nomads, who moves from one place to another (Kalsbeek, 1986; Kalton 2001; Kish et al., 1961; Kish, 1991).

Consider the case of studying persons visiting certain places, such as the health facilities, shopping centres, libraries or important religious functions in a community to study any characteristic of persons making the visits or some characteristic of the facilities that they visit. It is generally convenient to study them at the respective facilities. For random selection of the visits of persons at a location, no specific frame would be available.

The random selection can, however, be facilitated by the assumption that the visits occur randomly relative to the selection procedure. For example, one can divide the duration of the fieldwork into suitable time slots. Let us assume that the time slots are in half an hour duration, and a randomly selected slot is morning 10.00–10.30 am in a particular day. To obtain a random sample of visits, it is possible to randomly select a particular time in between the slot, say, 10.17 am and select the visit that occurs at this point of time.

Let the total number of visits that occur during the selected time period (10.00–10.30 a.m.) is also counted and is equal to $v$. The chance of selection of the visit can then be obtained as

$$\text{Probability of selection} = 1/v \tag{4.5}$$

The assumption is that those who visit during the time period are all equally likely to be included in a sample. This type of sampling is referred to as 'time–space or location sampling' or 'intercept point sampling'.

Before discussing the sample selection, it is necessary to have a clear idea about the objective of a study. Two types of objectives can be distinguished. Let us discuss this in the context of sampling persons visiting a shopping centre mall in a city. The objective of a study may relate to understanding socio-economic characteristics of persons visiting the mall or the reasons for their visit or the type of purchases intended or their opinion regarding any specific aspects of the services provided in the mall. In such cases, the visitors or the persons visiting the mall become the target population and one needs to take care of making the sample representative of the target population. There is no need to bother about those who may be having a zero chance of visiting the mall.

Assuming that the mall is a popular place to visit in a city, a second objective of a sample survey might be to study any other aspect of the city's population. For example, as against a household-based survey, a sample of visitors to the mall could be a viable and cost-effective alternative to understand the voting behaviour of the population. The target population, in this case, would be the population of the entire city and not those who would visit the mall. In this case, in addition to making the visits representative, one needs to consider whether every individual in the population has a non-zero chance of visiting the facility or not. If a sizable proportion of the population, particularly belonging to a specific group who are likely to have different political affiliation, has zero chance of visiting the facility, the sample estimate will have a bias and hence may not represent the population.

It is generally preferred, in a time-location survey to conduct the face-to-face interview at the facility itself. It becomes necessary in this case to consider: whether the place is suitable, length of the questionnaire, frame of mind and willingness of a respondent, etc. Generally, length of a questionnaire should be as short as possible and one should avoid asking questions that might embarrass a respondent in that context. There are doubts about reporting when people are interviewed in a public place about their sexual activities (Stueve et al., 2001).

### 4.4.1 Facilitating a cluster design

Let us discuss the sample selection approach considering only one facility, for example, a shopping centre mall in a city. If there are many such malls in the city, one needs an additional decision regarding how many malls to select and the process of their selection. The discussions provided in the earlier chapters will help in taking such a decision.

Before the sample selection, one needs to decide the point(s) where visitors to the mall are to be counted and their selection would be made. There are two obvious alternatives to decide about it. One is the point(s) of entrance/exit to the mall and the other any prominent place inside the mall.

If selection is done at a specific point(s) inside the mall, a person in a single visit can get multiple chances of selection, as many times as he crosses the selection point, during his stay at the mall. As already mentioned, probability of selection of a person in a visit depends on number of chances she/he gets for the selection. It is possible to count the number of persons crossing a selection point inside a mall or get an idea about how many times one crosses through a specified point by asking a relevant question to each selected person. But often it may become difficult to have such counting or get an appropriate reply, particularly if one has spent quite a long time in the mall. Unless there are administrative issues, such as opposition from the mall authority to conduct the operation at the entrance, it is better to consider the entry points rather than points inside the mall.

As a general guideline, it is advantageous to follow a cluster sampling design as indicated in Chapter 3. Clusters (PSUs) can be formed through a combination of location ($L$), entrances in a location ($E$), time ($T$) and number of days ($D$) for which the data collection process would continue. Total number of PSUs will usually be given by

$$\text{Total PSU} = L.\ E.\ T.\ D$$

In this example, since $L$ is assumed to be one, the total PSU will be a product of the last three factors. If unnecessary, some categories of a variable can be dropped. Let us suppose that a mall having four entrances ($E = 4$) remains open for 8 h/days except on Sundays when it works for only 4 h. Data collection is planned for 1 week. If half an hour duration is considered for selection purposes, $T$ will have 16 categories, except on Sunday when it will be 8. Hence, the total number of PSUs would be 416 ($4 \times 16 \times 6 + 4 \times 8$).

It may be mentioned that the duration of a study period ($D$) is directly correlated to the number of visits a person has. Hence shorter the duration, less it is likely to be the duplicate appearances by a person and also, easier for her/him to correctly recall how many visits she/he would be making within the study period. However, larger the duration better would be the sample in terms of giving chance to capture every segment of the population. For example, keeping the period as five working days in a week would mean having less chance (probably zero chance) of including those who have a heavy work schedule and prefer such visits only on holidays. Also, lesser the duration of a survey, smaller the number of interviews conducted per day by a team. One can of course increase the number of teams but having too many persons engaged in the data collection may mean inconvenience to others. Hence one has to decide an optimum duration taking into account these aspects.

For deciding a suitable time slot ($T$), following points may be considered. Too short a time slot may mean larger variation in the cluster size. Suppose, in a survey the time slot is 5 min.

A large group of people arriving together may swell the cluster size in one instance, whereas in another very few may enter. As mentioned, it is also better to count and note number of visitors entering in a given time slot. It is true that variation in cluster sizes can introduce sampling variance of an estimate. However, if $T$ is long, it will be difficult to count and there can be greater chances of error.

To clarify the procedure of sample selection that will usually be a cluster sampling, let us briefly recollect the salient points in the case of cluster sampling (Chapter 3). They are:

- Number of clusters should be as large as possible, which tantamount to keeping their size as small as possible.
- Clusters should preferably be selected using PPS sampling. If size of PSUs is not available, one should try and get an estimated size.
- If it is not possible to get their estimated size also, effort can be made to make the sizes approximately equal by adjusting the categories of the three variables, particularly by changing categories of time and days. For example, Saturdays and Sundays being holidays, one can expect more visits and hence consider a relatively smaller $T$ on these 2 days. For example, a time duration of 15 min for the holidays compared to 30 min on weekdays might help.
- It would be beneficial to stratify PSUs by their size.

The above guidelines can facilitate designing a two-stage sampling that will be close to drawing an EPSEM design and obtain a relevant ratio estimate and its sampling variance.

If required, initial observations can help in understanding the variation in the cluster size (detail discussion on such sampling is given in Sudman, 1980). Few visits initially can provide an idea about how number of people visiting the place changes during, say, busy hours and not so busy hours. Similar effort can be made to examine the differences in number of people entering the mall through the different entry points. Some of the points may be more convenient and hence are used more.

Formation of clusters will be more difficult if the objective is like studying people with HIV or the injecting drug users. One needs to spend sufficient time to understand the clustering of such people by interviewing knowledgeable people in the field. Consider the following example: in a study on female street prostitutes, a probability sample of about 1000 prostitutes was selected. The study constructed a spatial–temporal sampling frame of locations where the prostitutes are generally available (Kanouse et al., 1999). The locations were found to be streets where such activities were likely to occur. The list of locations was prepared after extensive consultations with knowledgeable persons. Clusters were then formed using three variables: area (locations of a street), day of the week and shifts (time when they usually worked).

### 4.4.2 Selection, Data Collection and Estimation

Once the cluster formation is complete, it is necessary to decide the number of clusters and visitors from each of them to be selected. It is true that both the volume of visitors and their socio-economic and demographic characteristics can vary and depend on all the four factors, $L$, $E$, $T$ and $D$. For example, there can be few entrances ($E$) through car parking. In the case where the facility is a library, a visitor in the office hours on a working day would be more likely to belong to elderly non-working population. Similarly, if one is considering the number of vehicles crossing through a toll plaza, it can heavily depend on both $T$ and $D$. It is, therefore advisable to stratify the clusters created by the four factors.

Apart from theoretical guidelines that more clusters and less number of visits from each selected cluster would be better, there are a few practical guidelines that need to be considered as well. Consider the following practical aspects: having too many investigators waiting at an entrance for data collection may not be feasible or once a person is selected, it will not be possible to keep him waiting for the interview. Considering a 30-min time slot, selecting around five visitors for a cluster appears adequate. In this case, assuming that the cluster refers to 10.30–11.00 am, one can decide to select the five visitors who enters at 10.30, 10.35, 10.40, 10.45 and 10.50 am. Apart from the investigators, there should be another person who would record the total number of entries during the period 10.30–11.00 am. The cluster selection will be done either by PPS (if size is available) or randomly. Each selected person will be asked to mention how many times he has visited or expected to visit during the fieldwork period.

Estimation can be done using the theory mentioned earlier, and the method of estimation will depend on whether stratification has been used. As discussed earlier, stratification of PSUs is desirable. For example, for selection in shopping centre mall, one can consider stratification of days into working days and holidays. Working days can have substrata according to time distinguishing working hours. If there is a separate entrance from car parking areas and the socio-economic characteristics of those using car parking entrances and other direct entrances differ, entrances can be considered as separate strata.

One hypothetical example is given below to demonstrate the computation of mean and sampling variance of a ratio estimate.

### Example 4.1

An EPSEM sample of 100 visitors to a city mall was selected in 20 clusters of five visits in each to obtain an estimate of monthly income of people visiting the mall. The data presented below (Table 4.2) provide monthly income (in thousands) and intended number of visits [cases with an asterik (*) indicate one additional visit intended and absence of it means no other visit intended].

'$f$' is the sampling fraction (=100/$N$), with $N$, which is unknown, referring to the number of persons visiting the mall during the survey. Since it is an EPSEM selection, the probability of selection of a person is $f$, if she/he would have only one visit during the survey, and will be equal to $2f$ for those who visited or intended to have an additional visit during the survey. For example, in cluster 1, all the five were having single visits and hence the sum of weights for it is $5/f$ (shown in column 8 of the table). The sum of weighted $y$ values in it (given in column 7 of the table) is $90/f$ [=$(20 + 25 + 14 + 13 + 18)/f$]. Similarly, for cluster 9,

$$\sum_1^5 w_i = (4/f + 1/2f) = 9/2f$$

TABLE 4.2 Monthly income and intended number of visits.

| S. No. of cluster | Monthly income (in thousands) for person $j$ in $i$th cluster, with $j =$ | | | | | $\sum_{j=1}^{5} w_i y_i$ | $\sum_{j=1}^{5} w_i$ |
|---|---|---|---|---|---|---|---|
| | 1 | 2 | 3 | 4 | 5 | | |
| 1 | 20 | 25 | 14 | 13 | 18 | $90/f$ | $5/f$ |
| 2 | 52* | 22 | 17 | 21 | 26 | $112/f$ | $9/2f$ |
| 3 | 14 | 19 | 16 | 23 | 24 | $96/f$ | $5/f$ |
| 4 | 10 | 12 | 56* | 18 | 60* | $98/f$ | $4/f$ |
| 5 | 9 | 21 | 45* | 62* | 28 | $111.5/f$ | $4/f$ |
| 6 | 15 | 22 | 20 | 55* | 62* | $115.5/f$ | $4/f$ |
| 7 | 62* | 25 | 15 | 72* | 28 | $135/f$ | $4/f$ |
| 8 | 29 | 65* | 9 | 25 | 64* | $127.5/f$ | $4/f$ |
| 9 | 12 | 24 | 16 | 55* | 13 | $92.5/f$ | $9/2f$ |
| 10 | 22 | 16 | 61* | 63* | 28 | $128/f$ | $4/f$ |
| 11 | 11 | 17 | 26 | 22 | 10 | $86/f$ | $5/f$ |
| 12 | 10 | 58* | 23 | 49* | 21 | $107.5/f$ | $4/f$ |
| 13 | 20 | 14 | 62* | 30 | 51* | $120.5/f$ | $4/f$ |
| 14 | 26 | 48* | 13 | 71* | 22 | $120.5/f$ | $4/f$ |
| 15 | 63* | 21 | 72* | 20 | 11 | $119.5/f$ | $4/f$ |
| 16 | 30 | 80* | 55* | 14 | 26 | $137.5/f$ | $4/f$ |
| 17 | 11 | 26 | 20 | 16 | 12 | $85/f$ | $5/f$ |
| 18 | 17 | 16 | 61* | 27 | 50* | $115.5/\text{f}$ | $4/f$ |
| 19 | 62* | 65* | 25 | 10 | 20 | $118.5/f$ | $4/f$ |
| 20 | 16 | 56* | 18 | 25 | 60* | $117/f$ | $4/f$ |

and

$$\sum_{1}^{5} w_i y_i = [(12+24+16+13)/f + 55/2f] = 92.5/f$$

Using equation 3.14

$$\text{the mean monthly income } (r) = \frac{\sum_{i}^{20} w_i y_i}{\sum_{i}^{20} w_i} = \frac{2233.5}{85} = 26.28$$

The numerator and denominator of the ratio are the sums of column 7 and column 8 of the table. Note that the value of '$f$' is not needed as it cancels out. Similarly, for obtaining variance of the ratio also, the value of '$f$' is not needed. Assuming $f$ as negligible and employing the formula given in equations 3.15–3.18, we obtain

$$v(y) = \frac{20}{19} \times \left( 254133.8 - \frac{2233.5^2}{20} \right) = 4955.41$$

$$v(x) = \frac{20}{19} \times \left( 364.5 - \frac{85^2}{20} \right) = 3.42$$

$$\text{cov}(yx) = \frac{20}{19} \times \left( 9393.25 - \frac{2233.5 \times 85}{20} \right) = -104.34$$

$$v(r) = \frac{1}{85^2} \times [4955.41 + 26.28^2 \times 3.42 - 2 \times 26.28 \times (-104.34)]$$
$$= 1.77$$

It may be mentioned that in the example, the response rate is assumed to be 100 per cent. In practice, this will not be so. However, that does not create any difficulty in the procedure to estimate the ratio and its variance. The non-response should be noted carefully.

## 4.5 HOUSEHOLD LISTING

A house listing operation provides the frame for selection of elements in selected clusters, in a multistage design. A listing operation facilitates random selection of units by giving chances of inclusion to each and every unit in a selected cluster. It thus lends support to ensure selection of a probability sampling, where all the units in a population get a non-zero chance of being included in a sample. It is important, therefore, in such an operation to enlist all the units with correct addresses in a given area, so that if selected a unit can be located and contacted. Generally, a household is the elementary unit considered in population-based surveys. Alternatively, dwelling units can also be considered in specific situations.

Listing operation involves preparation of a map of an area showing clearly its boundary as well as all households in the area. Households are generally listed in a specified geographic order. The pattern helps in minimizing the chance of omission of any of the households. It can also assist in reducing the effect of clustering. It is expected that the effect of clustering would be high among adjacent households or households that are very near to each other. The effect would gradually decrease with the distance. A systematic selection of required number of households from a list that is serially ordered should be helpful to capture the heterogeneity. A systematic sampling, therefore, is often preferred while drawing households using the prepared list.

House listing operation, however, involves cost and time. Therefore, in many surveys, particularly in rapid surveys preferences have been shown to use alternatives to detailed listing to cut cost. Immunization coverage surveys initiated by the World Health Organization is one such survey that has been widely used (WHO, 1991). Milligan et al., (2004), for example, provide details about the methodology and discussion on its advantages and possible disadvantages. The fact is that it is not, in the strict sense, a probability sampling.

There are alternatives to household listing that can be used to maintain the requirement of a probability sampling. Few such alternatives are discussed here. Their advantages and disadvantages are also indicated. It is, however, necessary to reiterate that a detail household listing should, as far as possible, not be avoided.

### 4.5.1 Two alternatives if listing is to be avoided

**Alternative 1**

Let us discuss this in the context of a two-stage design where PSUs are selected with PPS and a fixed number of households are to be selected from each selected PSUs without listing a selected PSU. Let us assume that:

$A$ = number of PSUs in a population
$H_i$ = number of households in ith PSU
$H = \sum_{i=1}^{A} H_i$ is the total number households in the population
$a$ = number of PSUs to be selected in a sample
$b$ = average number of households to be selected in $i$th PSU
$h = a \times b$, is the total number of sample households.

Each selected PSU would be divided into $k$ compact segments of approximately equal size, where

$$k = \frac{H_i}{b}$$

For example, if $b$ is equal to 20 households and $H_i$ is 300, it means the PSU will be divided into 15 segments each having approximately 20 households. One such segment will then be selected randomly and all households in the selected segments will be included in the sample.

Now,

$$f = \frac{h}{H} = \frac{a \times b}{H}$$

$f_1$ = probability of selection of $i$th PSU

$$= \frac{a \times H_i}{H}$$

$f_2$ = probability of selecting a segment in $i$th-selected PSU

$$= \frac{1}{k} = \frac{b}{H_i}$$

Since all the households, in a selected segment, are included in the sample, probability of selecting a household in a selected segment is 1.

Therefore, the probability of selecting jth household in *i*th PSU would be

$$= f_1 \times f_2 = \frac{a \times b}{H}$$

Hence, this will be a probability sample with an EPSEM design. The advantage is that it would not require any household listing. The interviewing team, after visiting a selected PSU, has to subdivide the PSU into $k$ segments of size approximately equal to $b$ and then include all households for interview in a randomly selected segment (for further details, see Turner et al., 1996). However, the method has few disadvantages. It is preferable to use administrative subdivisions of PSUs, if available, for segmentation. In this method, since the segment size depends on $b$, the likelihood of using administrative subdivisions can be ruled out. Creation of many segments, if PSU size is relatively large requires training and skill and hence involves some cost. Random selection of one segment often becomes difficult to implement. For example, there may be a tendency to avoid segments that are isolated and difficult to approach. This can, however, be taken care of by following certain instructions (Kish, 1995, p. 322–327): i) the segments should be numbered serially starting from a given location (e.g., northern most corner of the PSU), and ii) use of random numbers should be made serially from a list of such numbers supplied from the office. Another problem in this case is that households selected in a PSU are not scattered but belong to a compact segment which tends to increase the clustering and hence the design effect.

#### Alternative 2

Often PSU level frames are available, but may refer to a past date rendering them incomplete to some extent. If the degree of incompleteness is not a big issue, a linking procedure can help in taking care of the missing households (probably that are created recently). In this procedure, the available frames are still used for the selection of households in a selected PSU. However, the new or missing households are also given a chance of selection by linking them suitably with an existing unit. There are different approaches for the linking. One is that whenever interviewing team goes for collecting information from a selected household, say, $A$, it has to do an additional job of finding out if there is any household in the nearby areas that is not already listed. Suppose there is such a household, $X$, then a decision needs to be taken as to whether to include $X$ in the sample or not. Let $B$ be the next household (next to $A$) which is listed in the available list. The decision would be if the distance between $A$ and $X$ is less, compared to that of $B$ and $X$, then $X$ would be included in the sample, otherwise not. This, in effect, means that the newly created element, $X$ is attached with an existing element $A$, and its chance of occurrence is same as that of the element $A$. The procedure, therefore, gives a non-zero chance of selection to each and every element and hence reduces any possibility of biases on this account. This method is suitable when a list is available and the degree of incompleteness is not much. Also, this method is preferred if new households are scattered throughout the PSU. However, if the decision regarding whether $X$ is closer to $A$ is left to an interviewing team, it is more likely to be 'no' than 'yes'.

If a group of units clustered together, such as a hamlet, is missed, few different linking approaches are feasible to take care of it.

**(i)** After a quick enumeration of the units, they can be placed at the end of the available list. Let us discuss this with an example. Suppose there are 100 households, in a PSU, as per the available list and 20 are to be selected with equal probability, that is,

$$f_2 = \frac{20}{100}$$

For systematic selection, the interval will be 5, and if $R$ (random start) is 1, 20 units 1, 6,...,96 would have been selected. Now, if it is observed that a cluster of six households have come up recently, they can be numbered between 101 and 106. Continuing the selection, the additional selected units will be 101 and 106. If $R$ is between 2 and 5, only one additional unit will be selected. In this case, the additional units also receive a probability of selection equal to $f_2$.

**(ii)** Another option can be to attach all the missed units with the last element in the list; if the last element is selected then all the missed units are also included. That is, the probability of selection of a missed unit will be same as that of the last unit in the available list. In this case, though the design remains self-weighting, the variation in cluster size is likely to be more. Suppose a PSU is subdivided into smaller localities. In that case, the missed elements in a particular locality can be attached to the last element listed in that locality.

**(iii)** A third option could be to select $k$ units, instead of all the units, from the missed cluster. In this case, the probability of selecting a unit from the missed cluster would be given by

$$= f_2 \times \frac{k}{m}$$

where m is the number of missed unit. In this case, the design will not remain self-weighting.

## 4.6 KEY POINTS

- Two situations can arise while dealing with population consisting of two groups: i) persons with an attribute and ii) persons without the attribute. One is to estimate the prevalence of the attribute and the other is to study specific aspects of those with the attribute.
- To study persons with an attribute, the frame can be screened to eliminate those without the attribute (blanks) for the estimation. This increases the cost and there can be an error in screening if the attribute is a sensitive one.
- Another alternative is to work with the original frame and make the necessary adjustment at the time of estimation. The presence of blanks will tend to increase the error.

- If there are duplications (same element occurring more than once) in a frame, it can be handled in different ways. If feasible, the frame can be screened to remove multiple occurrences of the same units. Otherwise, necessary adjustment by estimating weights of each selected units can be made at the time of estimation.
- If a frame is incomplete and the degree of incompleteness cannot be ignored, an alternative is to use supplementary lists. Estimates from the different lists can be pooled suitably.
- Under many situations, no frame is likely to be available. This is true particularly in cases where the study unit, instead of referring to households, deal with visits made by persons, such as persons visiting library, health facility, etc. The random selection of visits, in such cases, can be facilitated by the assumption that the visits occur randomly relative to the selection procedure. The selection procedure can be random selection of appropriate time slots.

# Chapter 5

# Tackling Non-Sampling Errors

## 5.1 INTRODUCTION

The total error in a survey can be subdivided into two categories: a) sampling error and b) non-sampling error. The first type of error arises, as explained earlier, due to estimation that is based on a sample (not all the units) which is selected according to a pre-specified sampling design. The errors occurring due to any other reasons, including that which can arise at the time of implementing a sampling design and cause a sample estimate to deviate from the true population parameter, fall in the domain of the non-sampling error.

Unlike the sampling error, which can be estimated when a probability sampling is employed, it is difficult to estimate the extent of non-sampling error in a survey. That is, once it occurs, it is difficult to handle it. But there are ways to avoid the possibility of its occurrence. Also, for certain specific type of non-sampling errors, it is possible to make suitable adjustments at the time of estimation to minimise the effect of such errors. It is suggested that appropriate precautions should be taken at every stage, from the beginning to the end of a survey so as to avoid, as far as possible, the chance of occurrence of non-sampling errors.

This chapter focuses on the areas where the possibility of occurrence of non-sampling error is high, and enumerates how they are likely to occur and the steps that can be taken to minimise their occurrence. There are three areas, during the implementation of a survey design, where the possibility of such errors to occur is maximum. They are:

(i) **During sample implementation**: Error can occur if a sample as selected by a proposed design is not implemented in the field accordingly. This is referred to as the **coverage error.** Although occurring at a time of sample implementation, the resultant error is not included in sampling error but creates a non-sampling error.

(ii) **During data collection**: Once a sample is selected, responses need to be obtained from the selected individuals. Two types of error can occur at this stage. The first relates to the inability to obtain a correct response from a respondent. This is known as a **response error**. The other refers to the failure to obtain a response. Two different situations can lead to such a failure. One is where a selected respondent either could not be contacted due to various reasons or if contacted refuses to participate in the survey. This is referred to as a non-response and the error due to it as a **non-response error**. Another situation, related to non-response, is the item non-response. In this case, a respondent is contacted and information on most of the variables is sought except a few. That is, a questionnaire being canvassed is partially complete. Estimation of items (variables) on which information is missing can have an error similar to the non-response error. This is commonly known as **item non-response error.**

(iii) **During data processing**: Error can also occur while transferring data collected during a fieldwork to a computer for its analysis. If data are collected using a questionnaire manually with paper and pencil, it needs to be coded and entered into a computer. Entire process of coding and data entry requires utmost care and supervision to ensure that information collected in a field are entered properly. The chances of error in this process reduce if mini laptops, also known as computer-aided personal interview (CAPI), are used for the data collection.

In the following, we elaborate the coverage, non-response, item non-response and response errors in further details and suggest the possible preventive measures.

## 5.2 COVERAGE ERROR

The first step in a sampling design is to define the study population to be covered in the survey. Each and every unit in the population must have a non-zero chance of being included in the selected sample. This can get violated knowingly or unknowingly during the process of sample selection and its implementation in the field. Any such deviations can be a source of the non-sampling errors. For example, a common problem that often occurs in a household-based survey is about inclusion of military population and/or those staying in an institution. This difference, if estimates are based only on household population, need not be negligible in every situation.

Also, any difference between sampling units selected by a design and that which are actually implemented in the field can cause an error. Suppose, the sample design adopted is a two-stage selection consisting of villages and then households after listing the same. Any fault in identification of a selected village or in listing it for further selection of households can result in a discrepancy between a sample estimate and the population parameter. All these relate to a coverage error. In the following, a few such errors that generally occur in a survey, particularly a large-scale household-based survey, have been discussed.

### 5.2.1 Discrepancy between Study and Target Population

Let us consider a study on adolescent reproductive health. Often households having children of ages 12–18 years are considered to draw samples of the adolescents. This is likely to exclude those staying in institutions, such as boarding schools causing a deviation from the study objective. The anomaly can be removed by suitably modifying the objective and the target population. If size of the institutional population is not negligible and they need to be included, some remedial measures are necessary. A way out, in this case could be to conduct the study at a time when schools are likely to have vacations. This might help in minimizing the gap between the study and the target population. Alternatively, an effort should be made to supplement the household-based estimate by a sample of adolescents selected from the institutions.

### 5.2.2 Omission of Areas to Reduce Cost

A tolerable amount of coverage error is often accepted in order to reduce the cost of a survey. If there is a remote and an inaccessible area, there may be a need to weigh the possibility of removing it from the frame for sample selection. Following are the three points that need to be considered for the possible exclusion:

1. Proportion of the total population that would get excluded by the exclusion.
2. The difference, in terms of the characteristic being studied, between the population residing in the area and that of the rest of the population.
3. Cost of data collection in the area compared to that in the other areas.

The cost of data collection in PSUs from the remote area, if selected, is a major concern. If cost is not a consideration, the alternatives of excluding the PSUs should not arise. If characteristics of the excluded PSUs are expected to be similar to that of the total population (point 2), the question of their size (point 1) can be relaxed to some extent to save cost. The bias created by the exclusion is likely to be negligible.

Another way to tackle the issue, if the bias in excluding such PSUs is substantial, is to have them in a separate stratum and include them in the sample with lower probability. The balance can be restored at the time of estimation using appropriate weights.

### 5.2.3 Tackling Small PSUs

Another related issue is to deal with small PSUs. In rural India, villages are usually considered as the PSUs. There exist a large number of villages inhabited by few people. Such villages are often situated in remote areas and face problem of accessibility. Data collection from the small villages, if selected, can increase the cost. Socio-economic characteristics of inhabitants of small villages are generally different from the general population. Therefore, before taking any decision to exclude them from the frame, it is better to check that the excluded population is kept at a minimum (around 5 per cent of the total population).

Inclusion of small PSUs pose another difficulty. Assuming a two-stage design, a required number of samples are generally drawn from each selected PSU. Suppose, the size of sample in each cluster is 25 households. Any PSU smaller than 25 households, if selected, would lead to a loss in the sample size. Not only this, small size villages are more likely to decline in size in future. To guard against the loss in sample size, when village size used for their selection refers to a past period, a size less than 50 households (twice that of the sample take) is generally considered as vulnerable, that is, if selected, it might not yield the required sample take. The procedure usually adopted to deal with such a situation is as follows.

First, all PSUs of smaller than $k$ households are excluded from the frame for their selection. The value of $k$ is decided in such a way that the total exclusion remains around

5 per cent of the population. In the second step, assuming a sample take of 25 households, all villages having a size between $k$ and 50 households are linked with a nearby PSU. Following points are kept in mind while linking PSUs:

**(i)** each and every household residing in a PSU of size $k$ and above should get a chance of being selected in a sample, and
**(ii)** every PSU, whether having a single village or one or more linked villages, should have at least 50 households.

Suppose, a village of size 20 households is merged with a nearby village having 280 households. After linking, the combined village having the size of 300 households will now form as one PSU, both for the PSU selection as well as for selection of households at the second stage, if selected. The linking procedure should be done before the selection of PSUs. With the availability of fast computing machines, it should not be a problem to arrange PSUs in a geographical order to implement the linking procedure. This procedure is illustrated in Chapter 10.

### 5.2.4 Error in Identification of a PSU

Confusion can arise, leading to error in correct identification of a selected PSU. In the Indian context, two villages can have same name and hence can create confusion in the selection. Another problem could be that of bifurcation of a village. Let us discuss this with an example. A selected village $X$ was recorded (as per the last census) as having 1200 households. Subsequently, before the listing operation, it was bifurcated into two villages. One with around 800 households retained the original name ($X$) and the other with the remaining 400 households was given another name ($Z$). There is every likelihood that a listing team would reach the new village $X$ and list it remaining oblivious of the bifurcation and hence not including $Z$. A correct procedure would be to consider both the present villages $X$ and $Z$ and list all the households that existed within the boundary of the original $X$ as per the recent census, which was used for the selection of villages. If population of the bifurcated part $Z$ is not considered in the listing, its residents (around 400 households) would not get a chance of being included in the sample. This can bias the estimation.

### 5.2.5 Error in Segmentation of a PSU

It is customary to segment large size PSUs to reduce the burden of listing and hence cut the cost. A general recommendation is: to select one unit it is sufficient to list about 15–20 units. For hilly areas or difficult terrain, one can suitably modify it to listing 10–15 households for selection of one unit (Kish, 1995). Therefore, if sample take is 25 households, PSUs above 400 to 500 households can be segmented. In difficult areas, the limit could be PSUs of size 250 to 300 households.

Segmentation is performed either considering natural boundaries, such as a canal, road, pond, etc. Such segmentation has one advantage. It offers the possibility of making less error in identifying the boundary of a segment. However, sizes of such segments will vary.

Variations in size of segments need not be a problem. That is, it need not introduce variations in either the cluster size (number of units selected from a PSU) or the probability of selection of units. If sizes are known, the variation can be taken into account by selecting a segment by PPS sampling and then selecting with an EPSEM a fixed number of units (the sample take) from it.

If sizes of natural segments are not known, artificial segmentation would be preferred. In this case, segments of roughly equal size are created with the help of imaginary lines as the boundaries. The imaginary lines can be a cause for error in identifying the boundary of a selected segment.

Selection of one segment is difficult to implement in actual field. If the task of selection is left to investigators, there are chances that selection of segments that are less accessible or relatively larger in size (if natural segmentation is adopted) will be avoided. Such a tendency can introduce bias in an estimate. Apart from supervision, instruction needs to be given to arrange segments in a specific geographical order, and use random numbers, from a given list, to ensure strict randomness in the selection.

### 5.2.6 Error in listing a PSU/segment

Error or bias can occur if all units belonging to a selected PSU or a selected segment of it are not enumerated and listed. As indicated, chances are often high to exclude households that are situated in separate hamlets and in relatively inaccessible areas. There is another aspect that can lead to error if overlooked. This refers to cases where a person has more than one residence in a selected PSU. If such cases are not negligible, remedial measures are needed.

The problem can be taken care of at the selection stage, that is, at the time of listing. If a person has more than one house/household within the same PSU being listed, only one of them should be retained in the list and the others should be excluded from it to even out the chance of selection of households. The household to be included in the list can be decided by adopting a clear and uniform definition of a household member. Another way to compensate for the increased chances of selection (when a person has more than one household) is to use appropriate weights at the time of estimation. The listing and selection, in this case, can proceed as usual (listing all households belonging to one person).

## 5.3 NON-RESPONSE ERROR

Non-response error occurs when information on a selected unit is not available due to various reasons. If the non-response cases are not randomly distributed and have some specific characteristic related to a parameter being studied, it can create a bias in its estimation. In other words, apart from the size of non-response, the distribution of non-respondents according to characteristics that are related to the study variable will have implications for the error in estimation. The non-availability of response from a respondent, in a household-based survey, can occur in two steps. One, the household in which a respondent lives could not be contacted. The other is that though the household could be contacted, it is not possible to contact and obtain information from the respondent. To understand the implication of non-response it is necessary to compute the overall non-response rate, the non-response rate among persons. For this, let us define the following parameters:

$H_R$ = household response rate (proportion of eligible household from which response is available).

$H'_R = 1 - H_R$ = household non-response rate.

$EP_R$ = response rate among eligible persons (response obtained per eligible persons).

$EP'_R = 1 - EP_R$ = non-response rate among eligible persons.

$OEP_R$ = overall response rate among eligible persons.

$OEP'_R = 1 - OEP_R$ = overall non-response rate among eligible persons.

Therefore

$$
\begin{aligned}
OEP_R &= (H_R)(EP_R) \\
OEP'_R &= H'_R + H_R \cdot EP'_R = 1 - H_R + H_R - H_R\, EP_R \\
&= 1 - H_R\, EP_R
\end{aligned}
$$

For obtaining the household response rate ($H_R$), one requires the number of households that are selected for interview (the denominator) and the number of households from where responses could be collected (the numerator). Since the denominator is obtained from the household listing and the numerator from the actual fieldwork and there is generally a time gap between the two, an adjustment may be needed to obtain $H_R$. That is, some of the denominator may not qualify to remain as an eligible household. It can also be due to a mistake committed by a listing team. Let the total number of households selected as per the household listing be $H$, and

$H_1$ = number of households that could be located as well as interviewed,

$H_2$ = number of households located but interview could not be completed as no eligible members were present,

$H_3$ = number of households located but interview had to be postponed (partially complete),

$H_4$ = number of households that could be located but the interview was refused,

$H_5$ = number of households located but the interview could not be done due to any other reasons,

$H_6$ = number of cases where the dwelling in which the selected household was supposed to be located (as per the list) no household existed,

$H_7$ = number of cases where it was found to be an unoccupied dwelling,

$H_8$ = number of cases where the address was not a dwelling.

Then,

$$H = H_1 + H_2 + \cdots + H_8.$$

The appropriate denominator to compute the household response rate includes $H_1 + H_2 + \cdots + H_5$. The other three ($H_6$, $H_7$ and $H_8$) can be ignored because they do not indicate failure to obtain response, they are rather a problem either in the listing or due to a change that has occurred subsequently. The contribution of $H_6$, $H_7$ and $H_8$ should be small. The household response rate is

$$H_R = \frac{H_1}{H_1 + H_2 + \cdots + H_5} \tag{5.1}$$

Similarly, the results of non-response in obtaining information from eligible persons can be denoted as

$EP_1$ = number of eligible persons for whom interview is completed,
$EP_2$ = number of eligible persons for whom interview could not be completed (partially complete),
$EP_3$ = a person found eligible at the time of initial interview (to determine eligibility) turned out to be ineligible later,
$EP_4$ = number of eligible persons who could not be interviewed because of their absence,
$EP_5$ = number of eligible persons who refused,
$EP_6$ = number of cases where selected eligible person was found to be incapacitated,
$EP_7$ = number of cases where interview could not be completed due to any other reasons.

Suppose,

$$EP = EP_1 + EP_2 + \cdots + EP_7.$$

To obtain the response rate of eligible persons, the denominator, the number of eligible persons selected should exclude $EP_3$.

Hence

$$EP_R = \frac{EP_1}{EP - EP_3} \tag{5.2}$$

The overall response rate among eligible person ($OEP_R$), which is the product of equations 5.1 and 5.2, should be as high as possible. Generally, a non-response rate of around 10 per cent is acceptable. However, if the non-response cases can be considered as a random sample of the general population, the limit can be relaxed.

### 5.3.1 Remedy for Non-Response

As has been mentioned, the direct effect of non-response is that it reduces the sample size and hence influences the sampling variance of an estimate. The other effect, which could be more serious, is its biasing effect on the estimate. This occurs if the chance of occurrence of non-response varies according to some attribute of the units, which is also related to the characteristic being studied. To elaborate further, let us consider education as a characteristic that is related to both non-response and the parameter, extent of institutional delivery being studied. Suppose, education is positively related to both the extent of non-response and prevalence of institutional delivery. There will be a negative bias (underestimation) in the estimate of institutional delivery due to non-response. In the following, separate guidelines have been discussed to deal with the non-response if it is randomly distributed and in cases where it depends on the study characteristics.

#### 5.3.1.1 Adjustment for non-response when error is randomly distributed

Logically, to take care of a non-response, one can replace it with response of another unit which can be regarded as similar to the unit that did not respond. In fact, as has been mentioned in Chapters 2 and 3, the use of weights also works on a similar principle. Suppose,

$N_i$ = number of individuals in a population,
$n_i$ = number of individuals selected in a sample,
$r_i$ = proportion of selected individuals who ultimately responds and
$1 - r_i$ = non-response rate.

Assuming an EPSEM design, we have

$$\text{probability of selecting an individual} = \frac{n_i}{N_i}$$

$$\text{weight of an individual } (w_i') = \frac{N_i}{n_i}$$

The above weight is generally called a base weight, which is the inverse of the probability of selecting an individual specified by the design. It follows that:

$$\text{the probability that a selected individual will respond} = \frac{n_i r_i}{N_i}$$

Therefore, base weight adjusted for non-response $(w_i'') = \dfrac{N_i}{n_i r_i}$.

For example, if there are 200 individuals in a population and a sample of 10 are drawn with an EPSEM, the weight of each individual will be 20 (as if each individual is equivalent of 20 units in the population). If the response rate is 50 per cent, which means we fail to get information from five individuals, the adjusted weight will become 40.

It is not advisable and strongly recommended to not consider substitution to take care of non-response. It is often thought that it should be possible to consider a neighbouring household as a replacement to a non-responding unit. This can easily lead towards an unknown bias in an estimate. Following example might help to understand this.

In a survey to estimate the level of fertility (basically, to know the proportion of women who gave birth in 1 year prior to the survey) in rural India, a two-stage design was adopted to select villages (with PPS) and a fixed number of households in a selected village (systematically).

Substitution was allowed to replace a selected household with a neighbouring one, in the case of non-response. During estimation, the level of fertility was found to be unusually high. The reason for such upward bias was later found out to be because of the substitution. Since the fieldwork was done during a crop-cutting season, women (respondents) were generally outside in the field which resulted in a higher non-response. The substitution was selectively more among the women who have had a birth during last 1 year or so, because they were more likely to be at home to nurse the baby.

It is necessary, in case a substitution is allowed, to keep track of the details of substitution so that the probability of selection of each unit can be obtained. Even if a correct procedure is adopted to obtain the probability of inclusion of a responding unit, the procedure can unnecessarily complicate the weighting procedure and will increase the sampling variance of an estimate. It is, therefore, advisable to strictly avoid substitution.

#### 5.3.1.2 Adjustment for non-response when error is not completely random

The non-response, as mentioned, can be related to some characteristics of respondents. If known, it should be adjusted accordingly to avoid bias. A common situation prevalent in most household-based surveys in India is that the non-response rate tends to be higher in urban than in the rural areas. This can be easily taken care of by obtaining the corresponding response rates in the two areas. Suppose,

$N_r$ & $N_u$ = total population in rural and urban areas,
$n_r$ & $n_u$ = samples to be drawn in the two areas and
$r_r$ & $r_u$ = response rates in the two areas.

The base weights adjusted for non-response in the two areas (assuming an EPSEM design in both areas) will be:

$$\text{weights in rural} = \frac{N_r}{n_r r_r}$$

$$\text{weights in urban} = \frac{N_u}{n_u r_u}$$

This is a situation that is easy to handle. Although the weights differ among individuals in rural and urban areas, within a given area, they are same for all the responding individuals. Hence, it takes care of the biasing effect of the differential non-response between rural and urban areas in the combined estimate.

What if the non-response varies by level of education of a respondent? This, if not taken into account, can bias the combined estimate as well as the estimates in both rural and urban areas. One convenient way to consider this, as discussed and illustrated in Chapter 3, is to obtain the values of the response rates within each stratum, where stratification is done and adjusted accordingly.

The non-response can be high if every selected individual is contacted only once, that is, if no revisit is planned to consider those who were not available at the first attempt. Apart from increasing the magnitude of non-response, such an error can also introduce bias in an estimate. For example, in India, household surveys related to women can systematically have underrepresentation of working mothers and bias an estimate if revisits are not planned. This is because many of the characteristics of working women are different from that of the housewives (Ram and Roy, 2004).

Refusal is a category of non-response that is frequently related with the objective of a survey. For example, in a survey to estimate the 'hidden' income, persons having such income are more likely to refuse to be interviewed. If so, it is difficult to adjust it. The adjustment mentioned above need not rectify its effect. It is, therefore, important to keep refusals as low as possible. Revisits with more experienced investigators/supervisors should be planned to reduce the extent of refusals. If it still remains high, an alternative way is to draw a subsample of the total refusal cases and make a fresh attempt to obtain response from the selected subsample. Considering the refusal cases to be a separate stratum, the estimation can be done accordingly.

### 5.3.2 Item Non-Response Error

Item non-response is a special case where responses are available but for a few items (questions). They are commonly known as methods of estimating missing values or imputation method. One advantage in estimation of a missing value for a respondent is that a plenty of other characteristics of the respondent would be available and hence could be used fruitfully for the purpose of estimating the missing items (Brick and Kalton, 1996).

Let us assume that imputation is required for a variable $y$ in a survey. That is, while majority of the sample respondents have provided information on $y$, for some it was not available. Also, let education and place of residence (urban or rural) are two (for simplicity, we consider two but it can be more) independent variables that are related to $y$.

The aim behind imputation is to estimate a value that can replace $y$ for a respondent for whom the information is missing. Following three situations are discussed.

**(a)** The missing event $y$ is randomly distributed, and it does not depend on any independent variables. Also, the chance of missing is independent of $y$. Suppose, the event is the month of birth of a respondent. The respondent provides year of birth but the month of birth is missing. The month of birth is expected to be unrelated to any characteristic, such as education or type of residence. Also, the chance of missing does not depend on which particular month she/he was born. In such a case, a missing value can be substituted with the mean month of birth for all the respondents for whom the information is available. The calculated value is likely to be July. That is, for any respondent for whom the event is missing, it will be substituted by assuming that it will be the month of July.

If missing values are substantially large, substitution of a fixed mean value in each case will reduce the variability in the variable ($y$). It is possible to employ another method. Instead of substituting it by the mean value, one person from among those who have responded the event can be chosen randomly and his month of birth can be taken as the estimate for the missing value.

**(b)** The event $y$ is related to education and type of residence of a person. In other words, the value of $y$ varies according to education and type of residence but within an education and residence category, it is randomly distributed. The chance of missing like in the earlier case is independent of the event. Let the event be whether a woman, who has given a birth recently, received postnatal care. Whether the information would be missing for a woman would likely to be independent of whether she has had postnatal care or not, but the event is likely to be related to her level of education and residence. For estimating a missing value for a woman, her level of education and residence can be noted. A woman from the group whose response regarding the event is available and who belong to the same strata (education and residence category) can be chosen randomly and her response to the event will be considered as an estimate for the missing value. This is known as the hot deck imputation method (Andridge and Little, 2010).

**(c)** The missing event is related to $y$ itself. That is, the missing event is not random and depends on $y$. For example, if $y$ denotes the extent of 'hidden' income, those having a hidden income are more likely to avoid an answer. The estimation in this case is difficult. A number of techniques are available and newer methods are being added for estimating the item non-response (Rubin, 1987).

## 5.4 RESPONSE ERROR

Response errors can arise in different ways. The source of error can be from the respondent's side. The respondent may either not understand a question properly, or not have correct knowledge about the true response or knowingly provide an answer that is different from the true value. Very little can be done to mitigate it. The skill as well as training of interviewers can play a role in improving the participation as well as the response. Use of media to disseminate information about the importance of a survey and request for people's cooperation in the survey, if feasible, can improve the situation.

The source of error can be an investigator, who can pose the question in a wrong manner, may commit error in recording the reply unknowingly or knowingly, etc. An important source of such errors is the questionnaire itself. How a questionnaire is constructed is, therefore, important in reducing the response error. The other source of error is the investigator, his or her understanding of the questionnaire, interest, efficiency and integrity, etc. These issues are elaborated below. It needs to be noted that once non-sampling error occurs, it is difficult adjust it. It is always advisable to take every possible precaution to control its occurrence.

### 5.4.1 QUESTIONNAIRE CONSTRUCTION

The wording or framing of questions and the order of their placement are important and can help in reducing the response error. Although it is not practicable to cover much details on the art of questionnaire construction, few important aspects of it is provided in the following.

The broad classification of questions into two groups as factual and non-factual is beneficial in discussing their formulation as well as grasping the ensuing problems which can assist in framing questionnaire in a general situation (Sudman, 1983; Kalton and Schuman, 1982). The distinguishing feature of the two types is mainly the fact that whereas it is theoretically feasible to check the validity of a factual question, it is not so in the other case. For example, questions, such as age, income, education, health, etc. fall in the first group. The age of a person can be verified, for example, from the birth certificate or school leaving certificate. The attitudinal and behavioural questions belong to the other.

#### 5.4.1.1 Factual questions

While stating a question in simple and clear term is necessary irrespective of a type of question, it is particularly so for factual questions. This is because the clarity is often assumed in such cases. Let us consider, for example, questions on number of rooms and age. A query on number of rooms a household has though seems to be simple can have myriad of confusions. If not adequately supplemented by whether number of living rooms only or rooms including kitchen etc. are sought, the response to the question would depend on the subjective assumption (unknown) of a respondent as well as investigator and hence increase the variability. One needs to supplement the question, either by asking separate questions or giving further explanations, if possible, so that the confusions are eliminated. Regarding age of a person, though there seems to be little confusion, in certain situations (that particularly exist in rural India) it may be unknown to respondents. If so, and if it is left to the ability of either a respondent or an investigator to approximate, it can result not only in inflating the error but also introduce bias. Specific procedure is being used to train investigators to estimate age (discussed later) in such cases.

Often information on specific illness, birth or death of persons, expenditure or purchases made on different items, etc. are collected where a respondent is required to recollect an event. Such variables suffer from error that is generally known as recall lapse error. Considering a larger recall period results in one advantage. The chance of an event to occur ($p$) increases and hence the sample size required for it decreases. The recall lapse error, on the other hand, increases with the recall period. However, the relationship between duration (recall period) and the error in recalling it is not invariant and depends on particular events under consideration. For example, for recalling about minor illness, such as cold or fever, the recall lapse error is expected to increase rapidly with the duration, whereas for hospitalization with a major illness it is not so. It is for this reason the recall period is generally kept as 2 weeks for the minor illnesses, whereas it can be around 1 year for the latter. For births and deaths the recall period can be large, between 1 and 3 years. One point that should be noted in this context is though occurrence of an event, such as births, deaths, etc. is likely to be remembered, further details, such as whether a mother received postnatal care and by type of the care received would have higher recall lapse errors.

There is another dimension of the recall lapse error that is generally called 'telescoping'. It is the error in correctly placing an event within a given recall period. Such an error is repeatedly encountered in estimation of infant mortality, an important indicator, particularly of health in a population. Generally, information is collected on deaths of an infant among births occurring in, say, last 1 year. That is, whether a child born died in infancy and if yes whether it died at age 0 (less than one month), 1, ..., 11 months. In the case of a death, say, at an age of 11 or 13 months, a mother, who is generally the respondent, has to recall correctly whether it is within the reference period or not. Inclusion of greater proportion of infant deaths occurring at 13 months (or even 14 or higher) into the reference period than that of excluding those that occurred within the period will erroneously overestimate the mortality. If the opposite situation is true, the mortality will be underestimated. In the case of doubt, the investigator should try with further clues, such as trying to determine the month of death by asking about the season (like 'was it a rainy season?'), etc. Such issues should form part of training of investigators.

#### 5.4.1.2 Non-factual questions

To discuss the non-factual information, it is better to divide them into threatening and non-threatening types (Sudman, 1983; Kalton and Schuman, 1982). The distinction arises mainly in the case of behavioural data. Many behaviours, such as drug abuse, drinking, smoking, etc., which are considered as undesirable by a society are regarded as threatening type.

Similar is the case with information on persons having criminal record or the phenomenon of domestic violence, etc. In all such cases, the likelihood is that people will underreport such events. There are behaviours which are of the opposite type, in the sense that they are desirable and valued in a society. They are like working for some good cause, helping others in need, etc. Such behaviours are also considered as threatening because they are likely to be overestimated. Other types of behaviours, including attitudinal questions, are regarded as non-threatening.

It is difficult to avoid bias in estimating a threatening type of information. One possibility could be to frame the question little differently. For example, instead of asking a yes/no type of question, one can ask how often do you indulge in a specific behaviour and categorise the response as, say, regularly, sometimes, occasionally and never. Sudman (1983) suggested inclusion of another question to determine the degree of perceived threat of a behaviour. The question can be 'According to you, will asking about drug use make others feel very uncomfortable, somewhat uncomfortable, not at all uncomfortable?'. It can be categorised in more than three point scale. The advantage of this additional question while asking about the drug use is that the degree of threat felt by a person is likely to be directly related to the extent of underestimation or the overestimation. Thus, if in a population the perceived threat, revealed by the additional question, is low, the bias in the estimate is also expected to be less.

Randomized response technique is often used to estimate behaviours that carry stigma in a society (Warner, 1965).

There are few guidelines available for non-threatening or even factual questions. Consider, for example, ever use of family planning as well as its component parts, such as ever use of different specific family planning methods. The alternative ways to ascertain this can be: 1) to ask about the ever use, and if the response is 'yes' further enquire about the type of methods used, 2) ask about the ever use and along with it enumerate the names of all the methods, that is, an investigator would ask whether a respondent has ever used any method, such as naming serially the list of methods, sterilization, IUD, etc., and 3) ask separately for each method whether a respondent has ever used it. The method-wise estimation of per cent ever using is likely to be underestimated in the first two alternatives. Because there are several methods of family planning and a respondent, while responding to the first alternative is likely to miss a few. Reading a list of methods (alternative 2) also is not expected to help much because a respondent may not remember all the methods, as an investigator runs through the list of all the methods. It is expected that a respondent would report relatively more accurately the methods that are read either near the beginning or near the end compared to that which are read in the middle portion of the list.

It is generally better to be specific and ask about each method separately. The same is true if the objective is to ask about total expenditure during, say, hospitalization or about purchasing certain similar items from a grocery store. In the former case, it will be better to ask separately about expenditure on medicine, doctors' fee, etc. Often such expenditures are not known to a respondent. For example, if the expenditures are paid by insurance, employers or someone else. Necessary provisions should be made to deal with such alternatives.

Ordering of attitudinal type questions can often have important bearing on the extent of their prevalence. Consider following two questions in understanding the attitudes of people towards ragging in colleges in general and particularly about a sad episode where a student committed suicide because of violent ragging. The extent of unfavourable attitude towards ragging is likely to be higher if asked after the particular episode compared to that if it is asked before the narration of the other (Kalton and Schuman, 1982).

### 5.4.2 Errors due to Investigators

Investigators are the crucial link for obtaining information from respondents in a survey. The quality of data (responses collected from selected respondents) depends on their skills in understanding and delivering the required activities. It also depends considerably on the degree to which respondents were convinced and participated purposefully. In other words, soliciting cooperation and collection of relevant data from respondents also depend on the skills of individual investigators, how well they are able to garner such support. It is, therefore, natural to expect that there will be errors due to investigators.

Consider a hypothetical situation where two investigators are given a same sample (same respondents) and they carry out simultaneously the data collection work in an independent manner. If $\bar{y}_1$ and $\bar{y}_2$ denote the sample means for a characteristic obtained by the two investigators, the difference $(\bar{y}_1 - \bar{y}_2)$ will provide an idea about the error due to investigators. Ideally, the two estimates should be the same, which is likely to be rare. Hence, investigators bring in another component of variation in an estimate, and larger the number of investigators recruited in a survey greater is the expected variations or the error due to investigators. It is true that a survey, even with a sufficiently large sample size, can be completed fairly quickly by increasing number of investigators. But caution is required before advocating such an alternative. One needs to factor in the component of error due to investigators.

The above hypothetical example may be helpful in understanding the concept of error due to investigators, but it cannot be applied in practice. A practical alternative to understand errors due to investigators is to have few samples (of smaller size) drawn randomly and independently from a same population. These independent samples can be given to each investigator (or a team of investigators) to gauge the extent of errors due to investigators. Note that though we avoid interviewing same sample repeatedly by different investigators, the fact that they are given random samples from the same population justifies the expectation that any difference in the sample means, apart from random error that can be measured by the sampling variance, will be the error due to investigators. The moot point is each sample (subsample) is drawn from the same population and represents all the strata. Interpenetrating subsamples have been used, which were particularly geared to study errors due to investigators (Mahalanobis, 1946; Lahiri, 1958). One notable point is that since each sample is well spread throughout the entire geographic area of a population, the extent of travel by investigators increases, which can increase the cost of data collection. Also, though some idea about investigators' bias may be possible, it is difficult to incorporate it to make any adjustment in an estimate.

Is there any way to decide the number of investigators that would be suitable for a survey with a fixed sample size? The time frame planned for a survey is an important factor to decide it. A few other points are outlined below to help making such a decision:

- Larger the number of investigators, greater would be the difficulty in controlling their quality in terms of their education, experience in data collection, communication skills, etc.
- Training of investigators and supervision of fieldwork are the two extremely important components of a survey that helps in improving data quality. With increase in number of investigators, beyond a limit, it can hamper the quality of a training program. The extent of supervision also tends to get affected with the increase in their number.
- It is generally observed that the quality of work of an investigator, in terms of errors committed, is high at initial stages of data collection and it gradually declines as one gets more experienced. There is also a view that too much experience may also lead to higher error. In other words, there appears to be an optimum number of interviews per investigator.
- If the objective of a survey is also to provide an idea about the errors due to investigators, it would be better to allot sufficient number of interviews to an investigator so that a meaningful comparison of sample means of a characteristic for them can be made.

#### 5.4.2.1 Training

Training is crucial in controlling errors in a survey. Not that only the investigators need to be trained, everyone actively involved in a survey should participate in such training so that knowledge of all the basics are uniform among all concerned. Investigators and supervisors need to be trained thoroughly about the art of interviewing, steps to be taken to enhance the participation and interest of respondents, in addition to each and every question in the questionnaire.

Technical aspects of a survey instrument should necessarily be handled by a technical expert. For example, in a study of diseases, field staff should be taught about the basics of the disease. They should be given an adequate exposure to conducting interviews. Apart from mock interviews, they should, before beginning the fieldwork, individually interview sufficient number of cases as practice in the field.

#### 5.4.2.2 Supervision

Supervision is another extremely important component to improve data quality in a survey. Environment for good quality data collection starts from team composition. The process can be illustrated by discussing the arrangement followed in NFHS. In this large-scale survey with few thousand households as the sample size in a state, a field team consisted of one supervisor, one female editor and around three female investigators for the data collection from eligible women (there were few health personnel as well to assist in getting data on the clinical anthropometric and biochemical (CAB) components). Each investigator could complete an average of three to four interviews per day. Considering a cluster size of around 25 households (average number of eligible women per household was approximately equal to one), a team needed to stay in a PSU for 3 days.

The responsibility of a female editor was to check and edit each filled-in questionnaire, discuss with investigators about any mistake that they might be making and also observe an interview of the interviewers in her team. Before a team leaves a selected PSU, it is ensured that all the interviews are, as far as possible, completed as well as all the filled-in forms are edited. The supervisors, generally males, in addition to different administrative responsibilities, carry out some spot checks and help in filling the household schedule. Investigators share and gain from each other's experiences. Both the supervisor and the editor help in field supervision. In addition, supervisory visits were made by different other personnel responsible for carrying out the survey.

The performance of each team was monitored simultaneously with 'field check tables'. Such checks are particularly helpful in identifying the presence of certain distortion in data. For example, if the eligibility criteria for interviewing a female in a selected household are that she should be between the ages of 15 and 49 years, there can be a distortion in recording ages at the two endpoints, that is, at ages of women around 15 and around 49 years. There is a likelihood that women around 15 will be recorded as less than 15 years and similarly around 49 will be recorded as more than 49 years, in order to lessen the work of interviewing. Success of a survey in terms of controlling the non-sampling error depends largely on an effective training of investigators and supervision of the entire fieldwork.

## 5.5 KEY POINTS

- Non-sampling error can occur at different stages in the implementation of a survey design. Estimation of non-sampling error and its adjustment is difficult. Hence, it is prudent to take every precaution to check its occurrence.
- The coverage error mainly occurs at the time of implementing the sampling design. It can occur in identification of a selected PSU and/or in correct enumeration and listing of households in it. Employment of mappers and listers with appropriate and adequate qualification and training along with supervision of their work helps in mitigating the error.
- The non-response error occurs due to the failure of obtaining the required information from a selected element. Proper planning and sufficient training of the field team can help in reducing the error. It can also be controlled at the time of estimation of the parameters.
- Both respondents and investigators can be responsible for the occurrence of the response error. A respondent can commit an error in responding either due to lack of proper understanding of a question or knowingly. Framing of appropriate questions and pre-testing them in the field are helpful in decreasing the error. A thorough training of investigators and adequate supervision of their work can be helpful.

# Chapter 6

# Introduction to Evaluation Design

## 6.1 BACKGROUND

A research issue that is commonly encountered is to assess the effect of one variable (called treatment) on another, called an output. It is alternatively known as 'to measure or evaluate the effect of a treatment on an output variable', that is, to measure the change in an output created by the treatment. When a change in the output is exclusively due to treatment and not because of any other variable, it is termed as its causal effect. Evaluation designs facilitate understanding the causal effects.

A treatment, whose effect is to be determined, is generally considered to represent two categories denoted by the presence or absence of a characteristic. For obtaining a treatment effect, two sets of observations are needed, one having the 'presence of a treatment' (denoted as $T$) and the other called control would be 'without the treatment' (denoted as $C$).

It is not mandatory for a treatment to have only two categories. It is possible, for example, to measure the effect of more than one medicine for a disease in a single study. To gauge the effect of two medicines, we need at least three sets of observations on $C$, $T_1$ and $T_2$, where the first is the control (i.e., without any medicines) and the other two with the two medicines in evaluation. The focus can also be to comprehend the potency of a medicine or the optimum amount of a fertilizer to use. In this case too, one needs to have at least three categories, $C$, $T_1$ and $T_2$. Here, $T_1$ and $T_2$ would represent two different levels of potency for the medicine or the two different amounts of fertilizer.

As the variety of treatments to be included in a study widens, the scope for applying newer designs also arises. We, however, confine our discussion on development of designs to understand the causal effect of a treatment having only two categories, $T$ and $C$. A treatment can be a medicine or a vaccine used to understand its effect on a disease. The study can be designed to compare two groups of persons, one having the medicine or the vaccine ($T$) and the other consisting of those without it ($C$). It can be a particular variety of fertilizer in a study on agriculture to find out its effect on the yield of a crop by comparing the yield on two groups of plots, one with the fertilizer ($T$) and the other without it ($C$). A problem may be to know the effect of an awareness programme in increasing the demand for a product by comparing demand for the product in two groups, one exposed to the programme ($T$) and the other not exposed ($C$). A treatment can also be a public programme to examine the effect of providing cash incentives for increasing education among girls. The two categories of the treatment would be a community where the incentives are introduced ($T$) and the other with no such incentives ($C$).

The dependent or the output variables can be measured in an interval scale, such as monthly income, age, etc. or in ordinal scale, like degree of satisfaction from a programme measured in a five-point scale as 'very unhappy', 'unhappy', 'undecided', 'happy' and 'very happy' or else a binary variable, like whether feel happy, 'yes' or 'no'. A suitable indicator to examine a variable measured in an interval scale is mean or average, or a more convenient measure: the proportion (or per cent) or odds of individuals possessing a characteristic, in the latter two.

The study unit on which observations are made can vary from one study to another. In agriculture, the unit is generally a plot of land and the output is yield of crops. To understand the effect of an insecticide, leaves in trees can be the units of a study. In some cases, trees themselves can be considered as the unit. Also, even if the leaves are the unit, trees can be the sampling units. In the field of medicine, apart from human being, animals, like mice or pigs, are also often used as the study unit. In the field of social welfare or the programmatic research, the preferred unit is generally a group of individuals, like students in schools, individuals residing in a block or village as the study unit. There are implications of considering clusters or elements as the study unit. These are in terms of both the extent of matching that tends to control bias and the sampling error in a study (discussed later in Chapters 7 and 8).

## 6.2 BIAS AND ERROR

Suppose $y_T$ and $y_C$ denote the outputs corresponding to $T$ and $C$. Suppose,

$$t_i = y_{Ti} - y_{Ci}$$

which is the difference in outputs between $T$ and $C$ (the two treatment arms) in $i^{th}$ observation. $t_i$ can provide the effect of $T$ but not necessarily its causal effect. It can be a mere fluctuation that is random and regarded as error. The difference would show a systematic pattern in the presence of:

**(i)** causal effect of $T$ or the difference between the two treatments that occurs solely due to $T$, and/or

**(ii)** effect of disturbing variable(s), $D$, or the change between the two treatments that happens only due to $D$.

### 6.2.1 BIAS

The comparison will be biased if it ignores the effect of $D$ and regards the entire difference as the effect of $T$. In other words, the effect of $D$ is regarded as the bias present in an estimate. This is also referred to as the internal validity (Chapter 7). A design is an effort to generate data in a manner so that the distributions of $D$ are similar in the two groups,

$T$ and $C$. Because of the similarity in the distributions of $D$ in $T$ and $C$, the effect of $D$ is practically controlled. The outcome, $t_i$, which is because of either the treatment effect (causal effect) or error, becomes free of the biasing effect of the disturbing variables. This, in statistical terminology, is known as an experiment (for further discussion on experiments and surveys, see Kish, 2003b).

There are different types of designs that depend mainly on the following two aspects:

**(i)** How a comparison is being set up to obtain the value of $t_i$ (discussed in Chapter 7).
**(ii)** How the study units are allocated to the two comparison groups (discussed in Chapter 8).

The role of $D$ can occur in different ways. The diagrams below show how a change in $y$, an output variable, can occur, in the presence of $T$ and $D$.

Figures 6.1–6.3 show three different ways in which disturbing variables can influence and contribute to $t_i$ which unless eliminated will tend to bias the estimated causal effect of $T$. In Figure 6.1, $D$ has an effect on $y$, but it is independent of $T$. That is, $D$ does not have any association with $T$. The difference $t_i$ will indicate the effect of $T$, if any, as well as that of $D$. A design will, therefore, endeavour to balance the distribution of $D$ between $T$ and $C$ to remove the bias.

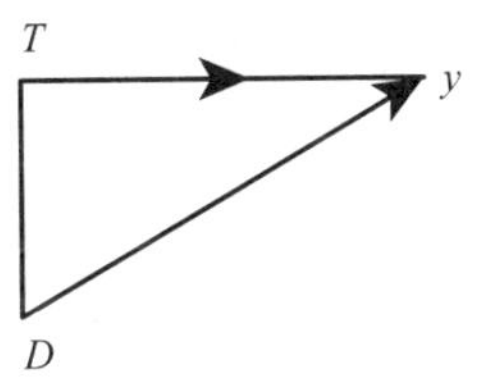


FIGURE 6.1 Effect of $D$ and $T$ on $y$ when $D$ and $T$ are unrelated

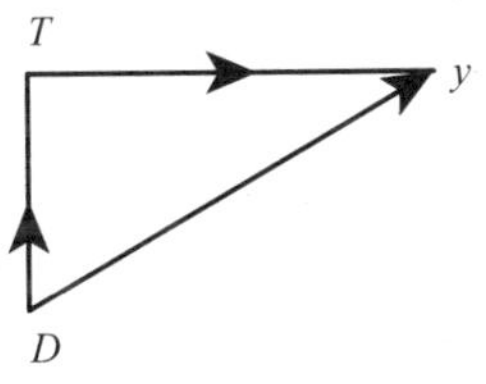


FIGURE 6.2 Effect of $D$ and $T$ on $y$ when $D$ influences $T$

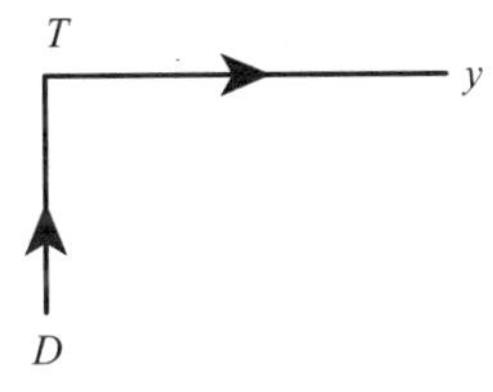


FIGURE 6.3 Effect of $T$ on $y$ when $D$ influences $T$ but not $y$

Let us clarify this with an example. Suppose, the objective of a study is to examine whether antitoxin is effective in enhancing the survival of patients suffering from certain disease (McKinlay, 1977; also discussed in Chapter 8). In other words, the purpose is to examine the effect of antitoxin ($T$) on the survival chance of patients ($y$): whether the administration of $T$ increases $y$? The patients are brought to a medical centre where they are administered the antitoxin. They are brought in different conditions, some with moderate symptoms and others in severe conditions. Naturally, the condition of a patient can be expected to be related to a patient's survival chance ($y$) and hence is regarded as $D$.

If the relationship is like that of the Figure 6.1, the difference $t_i$ will provide the effect of $T$, when a design balances, that is, makes the number of patients administered antitoxin ($T$) in moderate and severe conditions similar to those who were not administered antitoxin ($C$). It may be of further interest to also examine whether the effect of antitoxin changes according to the condition of a patient (in moderate or severe condition) at which it is being administered. This is known as the interaction effect between $T$ and $D$ in influencing $y$. This can be achieved through stratification discussed in Chapter 8 (also see Chapter 7 for the interaction effect).

Figure 6.2 indicates that $D$, in addition to influencing $y$, has an association with $T$. Unlike the interaction , the association means the value of $T$ (whether it will take a value $T$ or $C$) is dependent on $D$. Suppose in the example above, the variable $T$ is given a value '1' if antitoxin is assigned to a unit and '0' for those not given antitoxin or control. Similarly, $D$ is given a value '1' if a patient is in severe condition and '0' if in moderate condition. An association between $T$ and $D$ indicates that the antitoxin is not administered randomly, that is, the administration of the antitoxin depended on the condition of a patient. A positive association between the two means the chance of assigning antitoxin ($T = 1$) is higher among those who are in severe ($D = 1$) compared to moderate condition ($D = 0$). In this situation, a random sample of cases, if classified by $T$ and $D$, will show a higher proportion of it being administered in severe condition among $T$ than in $C$. Therefore, a simple comparison of proportion surviving among $T$ and $C$ will not show the causal effect of $T$.

However, if a design is used, which will ensure that the distribution of $D$ is similar in $T$ and $C$, the effect obtained can be unbiased and hence shows the causal effect. The association between $T$ and $D$ might make the design difficult to conduct because of a problem in getting the matched cases.

Figure 6.3 presents a situation similar to that in Figure 6.2, except for the fact that $D$ does not influence $y$. Here too the effect, if calculated by comparing the means (proportion of persons surviving) in $T$ and $C$ observed from a random sample of patients, will have a bias. The distribution of $D$ between $T$ and $C$ because of their association, are not likely to be similar in the two groups. In a design, when a distribution of $D$ is made similar between $T$ and $C$, the bias will be removed.

Another condition, apart from the one of controlling the effect of the disturbing variables, is about the temporal condition. It means that the cause should precede the effect.

Let us elaborate this with an example. Suppose, the objective of a study is to understand the impact of introducing a cash incentive scheme for accepting sterilization to increase use of family planning in a population. To understand the impact, one needs observations on two groups: one group referring to an area where the cash incentive (treatment) is introduced and the other without the incentive scheme. The effect can be measured in terms of the difference in the proportion accepting sterilization (the output) in the two populations. Apart from controlling all the disturbing variables, the output ($y_T - y_C$) must refer to the time when the treatment is introduced in the experiment. Suppose, the experiment started on January 1, 2012 when the scheme was implemented in the treatment group and its effect was to be computed after one year, that is, on December 31, 2012. It may be mentioned that sterilization (though reversible) is considered a one-time method and hence there is a difference between acceptance and use of sterilization. A person who accepted sterilization prior to 2012 can still be a user of the method in 2012. As per the temporal condition, the effect should refer to a comparison of proportion of acceptors of sterilization in 2012 (during January 31 to December 31, 2012) in the two groups. Bias can be introduced if comparison is made using sterilization users in the two groups in December 31, 2012. Similar situation might arise in the case of measuring the effect of a treatment (such as medicine, vaccine, any physical exercise, etc.) on a chronic illness, where the gap between incidence and prevalence can be substantial.

### 6.2.2 Bias elimination

There are different ways in which the bias can be controlled. The methods can be broadly classified into two groups:

- design-based approaches and
- model-based approaches.

The moot point in assessing a treatment effect is that it should exclude the influence of all the disturbing variables. The design-based approaches rely on the generation of data that is free of the biasing effects and would allow a controlled comparison to obtain the causal effect. One accepted approach to achieve it is to adopt 'random allocation' to distribute study units to $T$ and $C$ while generating the data. Random allocation means that the decision to assign a unit to $T$ or $C$ is taken by a mechanical procedure that provides equal chances to each unit to be in $T$ or $C$. This definition will be used throughout in this section (Chapters 6–9). It can be noted that the term random sampling refers to probability sampling, which need not be selection with equal probability. For selection with equal probability, the term EPSEM design has been used. There are other procedures, apart from random allocation, discussed in the next two chapters.

A model-based approach can also be used to achieve a similar purpose. To elaborate this, let us consider an example. In an athletic event, particularly, say, 400 m race conducted in a stadium, it is customary that the contestants do not stand side by side (horizontally) on a circular track, though their finishing point is the same. The person in the innermost circle stands first, person in the next circle stands at some distance ahead and it goes on like this. It is easy to understand why such an arrangement is made. The objective is to compare or measure the speed of contestants. This is done by comparing the time they take in running a distance. Since distance is a competing variable it is made equal for all contestants. A well-known model that relates distance travelled ($d$) with speed ($s$) and time ($t$) is

$$d = s \times t$$

This model indicates that for a given $d$, less the value of $t$ for a person, higher will be her/his speed. In a stadium, the things are already designed: markings are already made where the contestants should stand so that they run in equal distances. This allows one to judge the winner (one having the highest speed) as one who covers the distance first, that is, one who takes the least time to cover the same distance. What happens if the stadium is not well designed beforehand and the markings are not made? Is it still possible to decide the winner if contestants start at one place (on a horizontal line) and finish also at another place (on a horizontal line), on a circular track, so that they run different distances? That is, is it still feasible to achieve the objective, if generation of data is not designed but merely observed? An alternative possible way to decide the winner could be to take note of the distances that they actually run and use the above model to calculate the speed $\left(s = \frac{d}{t}\right)$. Since $t$ is also measured, the model-based analysis can help in making a controlled comparison. The model is based on certain assumptions, viz. uniform speed, similar fatigue, etc. and the generation of the data as done in the case of a design is not required. Such assumptions are avoided in the former case where the experiment is designed, that is, markings are done beforehand to control the distance. In the present example, the differences in the distance covered by persons standing in different circular tracks may not be much to worry about the assumptions.

There are well-known statistical models, such as regression analysis and covariance analysis, which are helpful in obtaining the effect of a variable on an output variable by simultaneously adjusting the effect of other so-called disturbing variables.

Often ethical issues may arise that can impede assignment of individuals randomly to a treatment, in a design-based approach. This can happen in every field but is particularly true in the case of medicine. Random assignment of individual raises the issue of imposing certain condition (beneficial or otherwise) to some who would be administered a treatment whose effect is being measured.

Also, under many situations random allocation of units is just not possible. Consider an example to understand whether obesity has a link with hypertension. In this case, *T* refers to the group consisting of obese and *C* a comparable group who are non-obese. Since only obese are to be considered in *T*, they are often self-selected. Alternatively, if there is a large pool of obese persons who agree to participate in the experiment, it is possible also to select randomly the required number of obese and such scheme will be better than the earlier one. However, this does not render it to be an experiment. What is required, in an experiment, is random assignment of a person to *T* and *C*, or that T and C are random samples from the same population. This is clearly not feasible in the present case. To design the experiment, one can try to ensure that the distribution of persons in *T* (who are obese) and *C* (non-obese) are similar in respect of all the disturbing variables. In such cases matching can be used, which can be pair matching or group matching (Chapter 8). Individuals in *C* can be selected by matching the disturbing variables with *T*. If the list of the disturbing variables is long or it is difficult to identify or control all of them, it may not be possible to control all of them. In such cases, the effect obtained is likely to be biased. The extent of bias will depend on two issues: i) how important is the disturbing variable(s) that could not be controlled, that is its relationship with the output variable being studied, and ii) the extent of mismatch or the extent to which the distribution of a disturbing variable differs between *T* and *C*.

The control of a few disturbing variables can also be exercised through statistical analysis while estimating the effect. That is, if a design is found to have not provided protection against a variable supposed to fall in the category of a disturbing variable, statistical analysis might be helpful in reducing its effect at the time of estimation.

Situation in causal effect analysis based on observational studies is different. Such studies are generally based on surveys, which do not pursue such internal balancing to have the causal effect. Let us consider a survey on employment. Let us also assume that information on education of a person is available from the survey. Can such data be used to obtain the causal effect of education on employment? Let us suppose that information on all the major disturbing variables is also available in the survey. However, since the objective of the survey is to represent the employment situation of a population, the survey data is not likely to have equal number of educated and uneducated persons. Also, question of balancing the disturbing variables will generally not arise, at the time of designing the survey. Multivariate analyses, such as regression and logistic regression, can be employed to facilitate obtaining the causal effect. One needs to check, however, the assumptions associated with these model-based approaches. There are also models proposed by Rubin that can provide the causal effect (Rosenbaum and Rubin, 1983, 1985; Rubin, 1974).

### 6.2.3 Error

As regards error, if all the disturbing variables are taken care of (either through the random allocation and/or matching), then under the null hypothesis that there is no effect of *T*, any difference between *T* and *C* that still exists, will be regarded as due to unexplained causes or an error. In the case of random allocation of units to *T* and *C*, the effect can be measured without any bias.

As shown by Fisher (1935), randomization can ensure a scientific investigation even when all the disturbing variables are not otherwise controlled (see Section 10.6.2). Randomization helps in putting all the disturbing variables into the category of random error (Kish, 1987). The size of the error in a design can be estimated (Chapter 7). The statistical tests to take care of the random error have been discussed in Chapter 9.

## 6.3 TYPES OF EVALUATION DESIGNS

Different terminologies have been used to cite designs for causal effects. Commonly, they are referred to as either experimental designs or evaluation designs. The need for establishing causal effect has existed since long, for example, to understand the effect of a medicine or vaccine. In these cases, the study units have been individual persons or animals. They have been termed as experimental designs. The issue of assessment of various welfare programmes, known as evaluation designs, is relatively recent. These programmes are usually assessed at a community or cluster level. Therefore, even though the control required to be exercised is similar in the two, the study units used are generally different in them. As discussed later, the extent of matching anticipation can be different in the two. However, if randomization is used for allocation of study units to treatment, the difference will only be in an increase in sampling variance of an estimate, if clusters are employed. In this book, we have used the term evaluation designs, but distinguish them by the types of control and study units used. The different types of evaluation designs are defined in the following.

### 6.3.1 Evaluation designs with random allocation of units

When study units are individual units and they are randomly allocated to *T* and *C*, the experiment is conventionally called 'true experiment' or 'experimental design'. We call them 'evaluation design with randomization of units'. The matching can be done, in such a design, before units are allocated randomly to *T* and *C*. The observations in an evaluation design with randomization are taken prospectively. For example, to understand the effect of a vaccine, medicine or a contraceptive, one can allocate persons (male or female or both as suitable) randomly to the two groups *T* and *C* and observe the appropriate output over a specified period of time.

Let us consider specifically the case of investigating the effect of a female contraceptive method in reducing the occurrence of accidental pregnancy. A random allocation of married women in reproductive age group of, say, 20– 44 years can be done (assuming that there is no ethical issues involved) to the two groups *T* and *C*. Suppose the observation period is 1 year. At the end of 1 year, a comparison of the pregnancies occurring among *T*, who were assigned to use the contraceptive and *C* who were not using any contraceptive would give the required effect. It is well known that fertility of women is related to their age and is higher among 20–29 years compared to 30–44 years. To make the design more efficient, matching can be done before the allocation separately within the two age groups.

Because of the random allocation of units to the two groups, this design provides complete protection against the occurrence of a bias due to the presence of any other disturbing variables. It is therefore regarded as the gold standard for the impact analysis. However, it tends to be weak in terms of the other two desirable characteristics, such as practicality and representation (Kish, 1987). In many situations, it is not feasible to employ such a design either because of ethical reasons or that random assignment is not possible. Since generally there would be restrictions on sample size, such a design also tends to be weak in terms of representation.

### 6.3.2 Evaluation designs with clusters allocated randomly

If clusters of units are randomly allocated to *T* and *C*, it is often referred to as the 'cluster randomized trials (CRT)'. We term the design as 'evaluation design with cluster randomization'. There is a need to distinguish between units for allocation and units for analysis. The PSU, in a study, may be clusters from which individual units are selected. The units for analysis may be individual or clusters or both, as deemed appropriate (Chapter 9). The distinction between this and the earlier design depends on the units for allocation.

It is clear that, irrespective of whether the unit is a cluster or a unit itself, their allocation to *T* and *C* is done randomly in the above two designs. The units before allocation can, however, be matched in terms of the major disturbing variables, though matching two clusters is not as exact as matching two individual units. Randomization is, therefore, a must and matching is optional.

### 6.3.3 Evaluation designs with unit level matching

In this case, the control is exercised only by matching units between *T* and *C*. Since randomization is not used, care should be taken in such designs to ensure that all disturbing variables, particularly the major ones are matched or are equally distributed in the two groups. These studies are often called 'quasi-experimental designs'. We refer them as 'evaluation designs with unit level matching'. For example, to examine whether obesity has a link with hypertension, one can have a group of individuals who are obese (*T*) and find a matched group *C*, who are similarly distributed as *T* in terms of all the disturbing variables but are themselves not obese. If there are genes that are responsible for both obesity as well as hypertension, such effects cannot be controlled. Both the groups *T* and *C* can be random samples independently drawn from a population of obese and non-obese. This may be helpful in terms of representation or generalization. However, the necessary protection required to estimate the effect or the difference in output between *T* and *C* is not provided unless the allocation of a unit to either *T* or *C* is done randomly, as in the case of the first two designs.

Since the random allocation is absent, the validity of the estimated effect will tend to become doubtful and makes the design weak, unless the possibility of biasing effects of other disturbing variables are taken care of. However, such designs have immense practical utility. Compared to evaluation designs with randomization, greater scope exists here to conduct an experiment.

One can use an alternative design to observe both *T* and *C* at more than one point in time to measure an effect. Such designs are helpful in reducing bias due to the presence of a disturbing variable (Chapter 7).

### 6.3.4 Evaluation Designs with Cluster Matching

In many situations either in business administration or programme evaluations, the assignment of treatment is often done at a cluster level rather than at an individual level (Chapter 7). In such cases, matching of clusters becomes the main focus. In all the other respects, these studies are similar to those that are discussed above. They are termed as 'operations research studies' or 'impact evaluation studies'. We denote them as 'evaluation designs with cluster matching'. Such designs are popularly used in evaluating impact of a programme.

### 6.3.5 Observational and Case-Control Studies

Strictly speaking neither observational studies nor case-control studies fall within the purview of evaluation design. In a case-control study, say, to understand the linkage between smoking and lung disease, one can get a number of persons having the disease (case) and those not having it (control) from a sample. One can then check in a retrospective manner the prevalence of smokers and non-smokers between the two. Such studies are popular in the field of epidemiology.

Already available data, particularly from a large-scale survey is generally utilized to understand the linkage between two variables of interest. Neither the allocation to treatments is done randomly nor any matching is done to ensure protection against the disturbing variables. An alternative that could be cost effective is to apply a model-based approach. A few such models are briefly discussed in Chapter 8.

Chapter 7

# Designs for Causal Effects: Setting Comparison Groups

## 7.1 INTRODUCTION

Different types of evaluation designs exist to study a cause-and-effect relationship. There are two requirements to create a design. One is to have comparison groups. The basic approach is to form two groups, one with and another without a treatment whose effect is being measured. These two treatment groups are denoted as treatment ($T$) for the one that receives it and the other as control ($C$) that remains without it. The basic idea is that units in the two groups are set up in such a manner so that their outputs can be compared for estimation of the causal effect of the treatment.

The formation of a design is complete when the requirement of creating a design, that is, assigning units to the two treatment arms, has been completed. This latter aspect provides the sanctity to the comparison to reveal the causal effect. The issue of assignment of units will be taken up in the next chapter. The present chapter elaborates the three basic comparisons that are widely used in the development of evaluation designs.

To choose a particular design in a given situation, one needs to consider the total error comprising both bias and error that is expected in estimating the effect under it, along with overall cost of the study. Apart from a description and estimation of the impact, in a design, the chapter includes discussion on the biasing effects that can arise and ways and means to handle them. It also specifies the computation of standard error of an estimate. In addition to a discussion on the comparison groups, the chapter provides elaboration on

- measuring main and interaction effects,
- bias and error in measurement of a treatment effect,
- sources of bias,
- internal and external validity and
- output and its timing.

## 7.2 MEASURING MAIN AND INTERACTION EFFECTS

Measurement of effect of a treatment needs at least two observations (rather two sets of observations) on a relevant output. It requires estimates of the output for *T* and *C* so that a comparison between the two estimates would provide the effect of treatment. The inherent assumption, in this comparison, is that the two estimates would be same if the treatment has no effect on the outcome of interest. In other words, the two samples providing the estimates are either drawn from the same population or selected in a manner that they do not differ in terms of factors, known as 'disturbing variables' or 'confounding factors', that could influence the outcome. The details of such allocations are discussed later.

There can be two different ways to set up a comparison. One is to have a distinct set of units for the two observations on *T* and *C*, and the other is to use a single set of units twice to obtain information on both *T* and *C*. In the latter case, it is natural to expect, since same units are used for both *T* and *C*, that the disturbing variables are automatically controlled in the comparison. However, it may not always be possible to get the two observations simultaneously. When there is a time gap in taking the two observations, the characteristics of sample units may also change with time and create an additional effect (discussed subsequently). For an ideal situation, therefore, one needs to have two observations on a same group of units that can be measured simultaneously.

Let us elaborate the point with an example. Suppose, a shoe company claims that their shoes with a new variety of sole (*A*) are more durable compared to another variety (*B*) existing in the market. To test whether *A* (the treatment) is more durable (measured by number of months a shoe lasts) than *B*, the control (absence of the treatment), it is possible to carry out an experiment in two different ways.

In the first case, one can set up an experiment by having two groups of persons: one designated as *T* wherein members will use sole *A*, and the other as *C* wherein members will use sole *B*. The members of two groups are observed till the shoe lasts. The question here is, whether the difference between the mean duration for which a shoe lasted in groups *T* and *C* is an indicator of the relative durability of *A* compared to *B*? The answer may be both yes and no. This simple difference indicator can be regarded as a valid indicator of relative durability of sole *A* compared to sole *B*, if and only if the groups *T* and *C* are similar in terms of the potential disturbing variables, such as number of hours a shoe is used for walking every day, type of road on which it is used, climatic conditions, etc. As these disturbing variables may directly or indirectly affect the duration for which a shoe can be used, if they are distributed differently among members of groups *T* and *C*, the overall difference in mean duration of use between the two groups will not reflect the 'true effect' of sole *A* on durability of the shoe. As for example, if members using *B* use it for more number of hours per day than members in *A*, it can lower the overall duration of use in group *C*, leading to overestimation of the effect of *A*. The estimated effect will be a correct measure of relative durability of sole *A* compared to sole *B* if it is possible to ensure that units in the two groups, *T* and *C*, are similarly distributed in terms of each of the disturbing variables. It may be noted that if the distribution of the disturbing variables are not similar in the two groups, the process will be known as 'observation' and not an 'experiment'.

An alternative way could be to set up an experiment in which each person uses a shoe made of sole *A* in one leg and another shoe made of sole *B* in the other. In other words, it is possible to get the two sets of observations on *T* and *C* simultaneously with a single group of persons. This can allow the experimenter to have complete control on all the disturbing variables, and hence to attain an unbiased estimate of the effect of *A*. Clearly, the second option would be preferred. However, such opportunity for control of disturbing variables may not be available in most situations.

Let us consider the example of estimating the effect of smoking on lung disease. In order to estimate the effect of smoking on lung disease, three approaches may be considered. Let us assume that the period of observation is 5 years. Ideally, one would like to have a single group of individuals and obtain two observations on the prevalence of lung disease in the group: one when they do not smoke, and the other when they smoke. The period of observation will be 5 years in both. Clearly, it is practically not feasible to have both the observations simultaneously. In another approach, if a single group is to be used, one can possibly observe the group for 5 years without anyone in the group smoking (*C*), and then observe the same group for the next five years when all of them smoke (*T*). Will the difference in the prevalence of lung disease in the two groups be due to smoking only? The answer is probably 'no'. Although the individuals are the same, the outcome for T is measured when the members of the group are 5 years older. It is known that age is itself a disturbing factor and is related to lung disease. Along with age, changes in lifestyle etc. may also be related with the disease. The difference in the prevalence of the disease between groups *C* and *T*, therefore, may not be solely due to smoking. A feasible third alternative in this case may be to obtain a group that is as similar to the present group in terms of the disturbing variables as possible, and designate one group as *C* and the other as *T* and observe them simultaneously for 5 years.

Whatever approach is taken in a particular situation (whether the two observations are taken on a single group simultaneously or after a gap of time or on two different groups observed simultaneously), the measurement of treatment effect, also called the main effect, can be computed as follows:

$$\text{Main effect of treatment} = y_1 - y_0 \tag{7.1}$$

where $y_1$ is the measure of output for *T*, and $y_0$, *C*.

Let us take two examples to further illustrate the procedure of estimating the main effect.

An experimenter is interested in assessing the effect of a fertilizer in increasing the yield of a crop. The experimenter can identify two sets of agricultural plots (*T* and *C*), which are as similar as possible in terms of the major disturbing variables, such as size of plots, soil fertility of the plots, amount of water used for the output, variety of seeds, etc. Any difference in these variables, for example, soil fertility in *T* and *C* can create disturbance in assessing the true effect of the treatment (fertilizer). If plots chosen for *T* have higher fertility compared to *C*, the experiment is likely to overestimate the effect of the fertilizer and vice versa. In other words, in the presence of disturbing variables the effect so obtained need not be the true effect of the treatment alone but may get mixed up with the effect of the disturbing variables. Now, let us assume that the plots are allocated to *T* and *C* in such a manner that the distribution of the plots in the two groups is similar in terms of all possible disturbing variables. The average yield ($y$) in $T$ and $C$ can then be obtained to get the main effect of the fertilizer using equation 7.1.

In another experiment, the interest may be to obtain the effect of a programme in increasing prevalence of institutional delivery where the programme is a door-to-door campaign on the benefits of the practice of having institutional delivery. Let us assume that the population consists of a congregation of villages. The experimenter can identify a group of villages that are similar in terms of the variables influencing the practice of having deliveries conducted at home or at institution (i.e., the disturbing variables). Such disturbing variables may be, in addition to some socio-economic and demographic variables, distance from the nearest place where facilities for institutional delivery exist, availability of road and transportation facility, etc. Let us also assume that each village is allocated to either $T$ or $C$ after controlling the disturbing variables. The campaign on institutional delivery can be conducted in only $T$ for, say, 2 years. The prevalence of institutional delivery ($y$) can then be estimated in $T$ and $C$ to get the main effect. Estimation in this case is likely to be done by selecting a sample of households where a delivery has taken place in the past 2 years.

It might be of interest, in some cases, to study the effects of two different treatments in a single experiment. Suppose, the experimenter is interested in measuring the effect of two drugs $A$ and $B$ for curing a certain disease, and it is also possible to use $A$ and $B$ in conjunction. The objective, apart from comparing the effect of the two drugs, is also to understand the effect when both are used together. In order to answer this question, we need to consider four groups, $C$, $T_1$, $T_2$ and $T_3$. Let $T_1$ be the first treatment (drug $A$), $T_2$ the second treatment (drug $B$), and $T_3$ is a combination of the two. $C$, as usual will be the control, not having any treatment. Let us denote the outputs (time taken to get cured) in the four groups as $y_0$, $y_1$, $y_2$ and $y_{12}$, respectively, where

$y_0$ = output in the control group,
$y_1$ = output in the group with first treatment,
$y_2$ = output in the group with second treatment and
$y_{12}$ = output in the group with both treatments 1 and 2.

As can be seen below, in this case, there will be two different main effects of each of the two drugs. The following equations state the two main effects of $T_1$.

$$\text{Main effect of treatment 1 (in the absence of treatment 2)} = y_1 - y_0 \tag{7.2}$$

and

$$\text{Main effect of treatment 1 (in the presence of treatment 2)} = y_{12} - y_2 \tag{7.3}$$

Note that in both equations 7.2 and 7.3, the comparison intends to show a difference that exists only because of the presence of the first treatment.

The overall main effect of treatment 1 can be computed as the average of the two effects in equations 7.2 and 7.3. That is,

$$\text{Main effect of treatment 1} = \frac{(y_{12} - y_2) + (y_1 - y_0)}{2} \tag{7.4}$$

Similarly,

$$\text{Main effect of treatment 2 (in the absence of treatment 1)} = y_2 - y_0 \quad (7.5)$$

$$\text{Main effect of treatment 2 (in the presence of treatment 1)} = y_{12} - y_1 \quad (7.6)$$

$$\text{Main effect of treatment 2} = \frac{(y_{12} - y_1) + (y_2 - y_0)}{2} \quad (7.7)$$

The computation of main effect of treatment 1 or treatment 2 (equations 7.4 and 7.7), however, assumes that the constituent effects (equations 7.2 and 7.3, and equations 7.5 and 7.6) are not different statistically. That is, any difference between the two main effects of a treatment can be regarded as having arisen because of the fact that they are based on samples (the details of the statistical test of significance are mentioned in Chapter 9). If this assumption becomes invalid, that is, if the two main effects of treatment 1 differ in the presence and absence of treatment 2, then it is said that there is an interaction between treatments 1 and 2. That is, the effect of treatment 1 varies and depends on whether treatment 2 is present or not.

The interaction between treatments 1 and 2 can be computed as the difference between the two effects of treatment 1, in the presence and absence of treatment 2 (similarly for treatment 2 in the presence and absence of treatment 1). Thus, interaction between treatments 1 and 2

$$= (y_{12} - y_2) - (y_1 - y_0) \quad (7.8)$$

or

$$(y_{12} - y_1) - (y_2 - y_0) \quad (7.9)$$

One should note that equations 7.8 and 7.9 are essentially the same equations written differently.

Let us assume the following values of $y$'s:

$y_0 = 35$ days,
$y_1 = 15$ days,
$y_2 = 18$ days,
$y_{12} = 8$ days.

Considering the absolute values of a difference, from equation 7.2, main effect of treatment 1 (in the absence of treatment 2) = $y_1 - y_0 = 20$. Similarly, from equation 7.3, main effect of treatment 1 (in the presence of treatment 2) = $y_{12} - y_2 = 10$.

If there were no interaction effect between treatments 1 and 2, these values would have been same statistically (i.e., even if there is a difference between the two, it arises because of the sampling fluctuations). The interaction effect is thus, from equations 7.8 and 7.9, equal to 10 days.

When interaction is present, the main effects lose their significance and are generally not computed. For example, if interaction is significant, it means that equations 7.2 and 7.3 (or 7.5 and 7.6) are significantly different from each other and hence a simple average of the two need not necessarily be the main effect of treatment 1 (or treatment 2).

It should also be noted that interaction is a phenomenon that can occur when there are more than one factor (cause) influencing an output (effect). The above example considers two treatments for the purpose of illustration of main and interaction effects. If there are three treatments, 1, 2 and 3, there will be three two-factor interactions (between 1 and 2, 1 and 3, and 2 and 3) and one three-factor interaction, between 1, 2 and 3 (for detail discussions on main effects and interactions, see Kempthorne, 1952; Namboodiri, 1970). Also, it is not necessary that interaction can be present between two (or more) treatments only. A treatment is a factor whose effect is being measured in an experiment. There can be different other factors; the disturbing variables in an experiment and, therefore, a variety of interactions are possible. Study of interactions, though important, require elaborate designs with large sample size that increases the cost of such experiments.

## 7.3 BIAS AND ERROR IN MEASUREMENT OF TREATMENT EFFECT

The main effects (equations 7.1 and 7.4) are the sample estimate of a treatment effect in a population from where the two samples, *T* and *C* have been selected, and one needs to verify whether the difference between *T* and *C* is statistically significant.

The essence of the test statistics that is employed for this purpose (discussed in Chapter 9) is that the expected value of the sample difference $y_1 - y_0$ is equal to the difference $\mu_1 - \mu_0$ that exists between the two population means. That is,

$$E\,(y_1 - y_0) = (\mu_1 - \mu_0)$$

where $\mu_1$ and $\mu_0$ are the population means in *T* and *C*, respectively. Under the assumption (null hypothesis) that there is no treatment effect ($\mu_1 = \mu_0$), the observed sample difference $(y_1 - y_0)$, if any, is assumed to have arisen due to the sampling fluctuations or an error due to sampling. This error, as has been mentioned in Chapter 2, can be estimated by the MSE or the standard error for an unbiased estimate.

It is important to note that the two samples from which $(y_1 - y_0)$ is computed can be considered, under the null hypothesis, to have come from same population. In other words, there is no biasing effect of any other variable and the difference is merely due to sampling fluctuations.

Consider the case of estimation of the effect of smoking on lung disease. Suppose, the sample for *T* is drawn from a population that is similar to *C* in every respect except for age distribution. The age distribution in *T* is younger compared to *C*. In such a situation, $(y_1 - y_0)$ is not expected to be zero under the null hypothesis. There will be a negative bias because lung disease is expected to be positively related to age, a disturbing variable that is not controlled (or similar in *T* and *C*). In the presence of a disturbing variable (i.e., if the distribution of the disturbing variable is not similar in *T* and *C*), the estimate of the treatment effect $(y_1 - y_0)$ will have a bias. Larger the presence of biasing effect of a variable, less will be the power of the test. That is, larger will be the Type II error, which means higher will be the possibility of getting a treatment effect when actually there is no treatment effect.

It should be clear that even if the two populations, *T* and *C*, are exactly similar in terms of all the disturbing variables, a sample from *T* might have relatively higher proportion of younger population compared to C and hence $y_1$ may differ from $y_0$. This difference will be considered an error and allowance is made for such errors by dividing the difference by standard error of the estimate ($y_1 - y_0$) (see Chapter 9). Larger the error (which depends on both bias as well as sampling variance), less will be the precision of the estimate, that is, more will be the Type I error, which means the null hypothesis of no treatment difference will not be rejected even if actually there is a treatment effect. It is, therefore, clear that the aim of a design to study cause and effect should be to control the overall error, particularly the bias.

To control bias, it is essential that either of the following two conditions is satisfied:

1. The units are allocated to *T* and *C* randomly.
2. The two groups, *T* and *C*, are free from the effect of any disturbing variables.

If, for some reasons, random allocation is not possible, then control of disturbing variables become paramount. In the absence of randomization, the presence of any disturbing variable will tend to bias the effect measure.

There are different types of comparisons that can be used in a design, each having certain advantages and disadvantages in terms of their suitability in reducing the bias in an experiment. Before discussing them, it is pertinent to understand the different types of bias that could possibly be present in an experiment.

### 7.3.1 Sources of Bias

It is possible to subdivide the bias into two major categories:

1. The two groups used for estimating the effect are different from each other in terms of the distribution of certain variables that can influence the output and hence can create a bias. These variables are termed as disturbing variables and their biasing effect is referred to as the 'selection effect'.
2. Another source of bias may be time: the 'time effect' that occurs when the same group is observed at two different points of time. The time gap can create a difference in certain variables between the two observations and hence can produce a biasing effect.

The effect of a variable can fall in a category of a selection effect or time effect, depending on the design being used for the impact evaluation. Let us consider the case of assessing the impact of smoking on lung disease. If *T* (smokers) and *C* (non-smokers) differ in age distribution, a comparison between the two will have the selection effect due to the age difference. If another design is used in which one group is observed twice (without and with smoking), their difference will have the time effect of age.

The selection effect of a variable can be controlled at the design stage (by making their distribution the same in *T* and *C*). The time effect is generally difficult to control at the design stage. They may be taken care of by making adjustment at the stage of analysis.

Disturbing variables are almost always present when two different groups are used for a comparison. In an experiment, it is necessary to make the distribution equal between them. While it is difficult to provide a list of such variables which will work in all situations, in many demographic as well as health-related studies, socio-economic and demographic characteristics, such as level of education, income, occupation, standard of living, age, sex and marital status are considered as the disturbing variables. The biasing effects of disturbing variables are controlled by suitably assigning units to the comparison groups. They are discussed in the next chapter.

Similarly, the time effects are generally present when the two observations on one group are taken at different points of time. The time effects are grouped into following seven categories:

**(a)** Unforeseen events: Some events, not directly related to an experiment, may occur during the course of information gathering that produces a bias in the measurement of treatment impact. As the name suggests, they occur suddenly and may occur anytime in between the two observations – before and after the treatment. In an agricultural experiment, suppose the yield of a crop (output) is observed at two different points of time without and with a fertilizer. A sudden rainfall may be such an unforeseen variable. It can influence the output observed at the latter time point and hence create a bias in estimation of the effect of fertilizer. Depending on when the event occurred and whether a rainfall is beneficial or harmful for the crop production, our estimate of effect of the fertilizer will be overestimated or underestimated. It is difficult to anticipate such events at the start of an experiment. Such confounding factors, also known as history and its effect, are called 'history effect'.

**(b)** Repeated exposure: A second type of time effect bias that might arise in an experiment that uses same units at two different points of time is mainly due to units' familiarity with the experiment. At the first time, a unit is likely to know the purpose of an experiment which can influence its response or reactions subsequently. This is also known as pre-test and the effect as 'pre-test effect' or simply testing and the effect as 'testing effect'. Clearly, such effects are likely to be present when human beings are the unit of a study and more prone to occur when psychology of individuals comes into play. For example, in studies on how people behave while participating in a demonstration, people's reaction is likely to be different if they are aware of being photographed, compared to when they are unaware of it. Familiarity can also influence measurement of a treatment effect.

Consider the case of evaluating a teaching programme on increasing intelligence quotient (IQ) of students. If an IQ test is administered on a batch of students before and after introduction of the programme, the difference between the mean IQ score of students before and after the programme need not be exclusively because of the teaching programme. Even without any such programme, the mean IQ obtained second time is likely to be higher than the first by five points (Campbell, 1957). Therefore, treatment effect, calculated on the basis of equation 7.1, without considering the effect of repeated exposure to the IQ test is likely to be biased (overestimated).

Consider another example of a laboratory experiment that was done to measure the impact of an advertisement on cinema/television on individual preference to use a particular brand of shampoo. The individuals agreeing to participate were told that there will be a drawing and whoever wins will get free shampoo worth an attractive amount. Participants' initial preference on brand of shampoo was noted. They were then shown a 30-min movie. In between the movie, three advertisements were shown, one of them being that of the particular brand of shampoo. At the end, their preference for the shampoo brand was noted again (Boyd et al., 1989). Can the difference between the observations on per cent preferring the particular brand of shampoo before and after the movie be considered as the effect of the advertisement?

Although individuals were not told about the purpose of the study, since they were asked about their preferences of shampoo before the movie, some smart individuals might notice the advertisement of the shampoo and respond accordingly when asked about the brand later. The pre-test effect can create a serious bias in many situations unless a conscious effort is made to control it.

**(c)** Maturation: With time, changes occur among people. While their age necessarily increases, changes among individuals may occur in education, occupation and exposure to outside world, etc., and these changes may affect their behaviour, knowledge, attitude, practices, etc. Some of the effects measured through equation 7.1 might be a result of such changes, and, if so, can disturb the validity of an experiment. Such biasing effects are known as 'maturation effect'.

**(d)** Measurement: Measurements are required to collect information on outputs. If equipment used for the measurement changes at two points of time, it may become a source of change in the output and hence bias the treatment effect. Consider an experiment to understand the effect of a particular type of food on level of anaemia among women. The level of anaemia, which is characterized by a low level of haemoglobin in the blood, can be measured using different techniques, such as using HemoCue machine or the standard blood cell counter (BCC) and it is believed that a measurement can differ depending upon the particular technique that is being used. Hence, if the measurements of anaemia before and after the nutrition intervention at two points of time are obtained using two different measurement techniques, it can create a biasing effect. Even if the same equipment is used at two points of time, with use it may decay and as a result may produce a difference in the measurements at two points of time. Similarly, bias may also arise when human beings are employed to observe and record a particular measurement or record it through personal interview. Even if the same person is recording observations at the two points in time, difference may occur because the person becomes more experienced over time. Bias can also occur if different types of tests, such as IQ tests, are used at two different points of time. These biasing effects can be grouped as 'measurement effect'.

**(e)** Dropout: When the same individuals are used at two different points of time (i.e., a panel), it is very likely that there will be some dropouts at the time of the second observation. This can be due to migration, mortality, failure to participate in the programme, etc. Although dropouts may as such mean only a reduction in the sample size and hence an increase in the sampling variance of the estimate, it can also create a bias. A biasing effect can arise if the dropout is non-random and tends to occur selectively among persons with certain characteristics. For example, mortality is likely to be age selective, chances of its occurrence being higher at very early and older ages compared to the young adult age group. Migration may also be age selective. Persons in the age group of 25–59 years are more likely to migrate and hence become dropouts in comparison to the other age groups. Hence, a sample that was drawn, say, with an EPSEM (equal probability selection method) design need not remain so after a lapse of time if there are selective dropouts. Resulting bias due to the effect of such changes are referred to as the 'dropout effect'.

(f) Multiple treatments: There is another effect that might arise if the same unit is given more than one treatment with some time gap. The effect may happen because of the presence of some left over effect of an earlier treatment. This is referred to as the 'multiple treatment effect'.

(g) Regression: If the same sample is used in both baseline and end line, in a before-after design (detailed in section 7.4), for evaluation, bias can occur if selection is not completely at random but pertain to a specific section of the population. For example, for evaluating the impact of a teaching or training programme, selection may consider those who are either 'poor' or 'good' performer. This bias, known as the 'regression effect', refers to a tendency of individuals to regress towards the mean. For example, if evaluation of a teaching programme is conducted considering all good students, a before-after design with the same group may result in an underestimation of the impact. A good student, either because of complacency or lack of necessary drive may fail to score high to show the real impact. The reverse situation may occur if selection is from poor students only. Because of enthusiasm or special effort from poor students, the impact may be overestimated. It is better to avoid such selection.

### 7.3.2 Internal and External Validity

Internal and external validity are two important concepts in bias in measurement of treatment effect. A treatment effect measured in an unbiased manner is known as internally valid. The validity refers to the absence of any biasing effect, such as the selection effect, effects of history, pre-test, maturation, etc., which can vitiate the estimation of a treatment effect. Since treatment effect is attributed to a comparison of two similar groups, $T$ and $C$, it is known as internal validity.

The external validity, on the other hand, is about the generalizability of a treatment effect. That is, to what extent the treatment effect obtained through a specific sample could be generalized or replicated for the population. This issue is therefore related particularly to the manner in which a sample of the treatment group ($T$) has been drawn. Since the role of control group C is mainly to ensure internal validity, the issue of external validity would depend more on the manner in which the treatment group, $T$, is selected.

Let us discuss this in light of the example to determine the effect of smoking on prevalence of lung disease. Suppose, we draw a random sample (of adequate size) of persons from a city in which the experiment is being conducted. Also all of them agree to be smokers and will be in $T$. A comparable control group, $C$ is then obtained for measuring the effect. This effect can be generalized for the city as a whole, because $T$ is a random sample drawn from the city. But will it hold good for whole of the state to which the city belongs or the whole country? Although the result can be extrapolated with appropriate weights (see Section 2.3.3), the answer will depend on whether $T$ can be considered as a random sample from the state or the country as a whole.

Both internal and external validity serve important purposes. The former provides credibility to the measurement of the effect, and the latter tends to assure that an effect obtained would be generalized or replicable for a wider population. To take adequate care of both in a single experiment is quite tricky. Since the estimation of treatment effect comes first, the internal validity is of prime and paramount interest. A pertinent question can be to what extent an experiment should attach importance to the external validity?

While external validity may not be of supreme importance to consider in some situations, e.g., in field of medicine, it is different for experiments in the field of marketing or programme evaluation. Consider the following two problems: one, regarding the effect of a particular type of advertisement on the sale of a product, and the other on door-to-door campaign on institutional delivery.

An important aspect of these studies is often to throw light on whether they can be upscaled or generalized for a larger population. This is true that the question of internal validity remains important but the external validity is also an issue that requires adequate consideration in such cases. Establishing the effectiveness of a new medicine is itself a big achievement. The question of its validity in different population can be taken up separately. However, it seems not enough to say that an advertisement will have a positive effect on sale, unless there is an adequate indication about its generalization.

## 7.4 THREE BASIC DESIGNS FOR ESTIMATING TREATMENT EFFECT

The basic requirement of setting up a comparison group is to have two sets of units (samples) to observe the outputs for the treatments, $T$ and $C$. Ideally, one would like to utilize same individual units for both the outputs and observe them at one point in time to take care of all the biasing effects. However, in several practical situations it may not be possible to achieve this. In the following, three basic comparisons have been outlined.

### 7.4.1 One Sample Each from *T* and *C* at Two Different Times (Before-After Design)

The basic feature of this design is that the two observations are made not at one point of time but after a gap. The procedure of estimation of the effect and the presence of biasing effects and how to control the biasing effects have been discussed below. Finally, the estimation of standard error of an estimate, which helps establishing whether an effect is a causal effect or an error, has also been provided.

#### 7.4.1.1 Description and estimation of effect

The before-after design is particularly useful for evaluating the impact of an intervention program. Usually, the initial observation is without the treatment (i.e., the control group), denoted by $C$. After the initial measurements are taken, the treatment, generally an intervention, such as a teaching or a training or an awareness programme, any advertising or marketing strategy, any programmatic intervention like an incentive programme, etc.

is introduced. After a gap of time considered appropriate to examine the impact of the treatment, the second observation, $T$, is carried out. If survey is the means of collecting the two observations, the initial observation is generally known as the baseline survey providing relevant information on $C$, and information on $T$ is obtained by an end-line survey. It may be mentioned that more than two observations can further strengthen the evaluation of an impact. Similarly, it is possible to examine the impact of more than one treatment simultaneously in an experiment (Kish, 1987). However, in the present book we consider evaluation of a single treatment with observations at two points in time.

The groups that are observed at two points in time (baseline as well as the end-line surveys) can be the same set of individuals (a panel) or they can be two independent samples from the same population under study.

A panel study has certain analytical advantages. It can provide better idea about the cohort changes. The method of selecting two independent samples has a major advantage of being free from the effect of the repeated exposure. Also, if the focus of a study is to envisage the increase due to a treatment and not the detail cohort changes, the two distinct samples will be advantageous.

A third option can be to have a mix of the two, that is, the total sample size at the end line can be subdivided into two parts, one being a portion of the initial sample enabling to analyse the cohort effect and the other an independent end-line survey.

Let us assume that $n$ denotes the number of units selected at both points in time and $y_0$ and $y_1$ are the measures of the study outputs on $C$ and $T$, respectively. The estimate of treatment effect ($I$) is obtained as

$$I = y_1 - y_0$$

#### 7.4.1.2 Biasing effects and remedies

This design has been a popular one in evaluation of treatment effects. A strong advantage of this design is that the disturbing variables are likely to be automatically controlled irrespective of whether a cohort approach is taken or two independent samples from the same population are considered. As a result, in cases where the time gap between the two observations need not be large (to control the history effect) and the pre-test effect is either absent or negligible or can be controlled, this design can be a very potent design in estimation of a treatment effect. Possible biasing effects of the different factors related to time and possible remedies, if any, to reduce the effects are discussed below.

**(i)** Bias due to pre-test effect and possibilities to reduce it

Unlike the other effects related to time, the pre-test effect ($P$) seems to be invariant of the time gap between the two observations. The magnitude of the pre-test effect of course can vary with time. Since it arises because of the familiarity with the programme, in many situations, the magnitude of the effect is likely to be inversely proportional to the time gap. Consider the case of measuring the effect of a teaching programme on IQ level of students mentioned earlier. The sooner the second test of IQ is taken, the greater is the possibility that a student would remember about his experiences in the earlier test and the effect would be stronger.

It is clear that the before-after design in which the same individuals are selected twice should be avoided if the pre-test effect is likely to be large. The effect $P$ has been a major concern for this design. It is, however, possible to reduce the effect of $P$. This can be illustrated with an example.

A study was conducted in the city of Cincinnati, OH, USA to measure the effect of an experimental campaign of information in increasing mass education and also peoples' attitude towards the United Nations (as reported in Campbell, 1957). Without going into detail, an attempt is made here to highlight the procedure used to reduce the biasing effect of $P$. The treatment or intervention was an intensive publicity campaign using different mass media of communication.

Two random samples from the city of Cincinnati were selected and the interviews were conducted in one of the two samples before introduction of the intervention. The intervention was carried out, and after 6 months interviews were conducted in the second sample for the study. It may be mentioned that the cohort interviewed earlier were also interviewed at the time of the end line by another organization. Hence, the following information is available:

$y_0$ = estimate of the parameter (say, per cent aware about any important happenings included in the intervention) from first sample before introduction of the intervention,

$y_1$ = estimate of the parameter from second sample after intervention,

$y_2$ = estimate of the parameter from the first sample after intervention

Estimate of the treatment effect was computed as

$$\text{Treatment effect (when } P \text{ is not present)} = y_1 - y_0 \qquad (7.10)$$

It can be seen that the treatment effect, in this case, does not consider $P$ because $P$ is not present. $y_1$, the observation after the intervention refers to the second sample that was interviewed for the first time and there is no repeated exposure. Can this be considered as before-after design, in which one group is not observed twice for the purpose of evaluation? The answer is probably yes, because they refer to two random samples from the same population. The time gap of 6 months in observing the two samples can be ignored. The study found that there was hardly any improvement in the awareness level. In other words, there was no treatment effect.

The study also revealed a significant effect of $P$. The treatment effect along with the biasing effect of $P$ was calculated as

$$\text{Treatment effect ignoring the effect of } P = y_2 - y_0 \qquad (7.11)$$

The difference in the awareness level when same group is interviewed twice (equation 7.11) revealed a very large effect of the treatment because of the presence of the biasing effect of $P$ that shows the vulnerability of impact estimation if same individuals are interviewed twice. Hence, if pre-test effect is known to be negligible, in a study, a cohort approach may be preferred. Otherwise, it would be advisable to have two independent samples from the same population.

Consider conducting the evaluation with two similar samples. That is, obtain two independently and randomly selected samples from a population, one at the start, that is, before introduction of the intervention and the other at the end when the impact is to be evaluated. The individuals selected in the end-line survey, since selected independently from the same population, may have some who were also selected in the baseline survey. Let $n_1$ and $n_2$ be the sample sizes at the baseline and the end-line surveys and $m$ be the number of individuals among $n_2$ who were also selected in the baseline. It is possible to have the following estimates:

$y'_0$ = estimate of the parameter based on $n_1$
$y'_1$ = estimate of the parameter based on $n_2 - m$
$y'_2$ = estimate of the parameter based on $m$ individuals
$y'_3$ = estimate of the parameter based on $n_2$

The treatment effect can be estimated as

$$\text{Treatment effect (without the effect of } P) = y'_1 - y'_0 \qquad (7.12)$$

The magnitude of $m$, which depends on $n_1$, $n_2$ and $N$, the population size, will generally be negligible. Suppose, 0.5 per cent of the population is selected at base line. A sample of size 1000 at the end line is expected to have only five individuals as common in both the surveys. Therefore, for all practical purposes an estimate based on $n_2$ and $n_1$, that is, $(y'_3 - y'_0)$, will be adequate to have the treatment effect (without the effect of $P$). If $m$ is sufficiently large, the treatment effect can also be computed as

$$\text{Treatment effect (with the effect of } P) = y'_2 - y'_0 \qquad (7.13)$$

Equation 7.13 will provide an idea about the effect of $P$. If this is not significantly different from equation 7.12, simple average of the two estimates can be taken as the treatment effect. If they are significantly different from each other, that is, there is significant effect of $P$, equation 7.12 can be accepted as the treatment effect, provided $(n_2-m)$ is sufficiently large.

There could be a slightly different type of effect that is close to the pre-test effect. This effect occurs because of a psychological effect of participating as a unit of a treatment group. This is often true in experiments to estimate impact of a medicine. For example, consider designing a study to find whether a sleeping pill is effective in increasing number of hours of sleep. The treatment group can be a group of randomly selected persons willing to participate. The output, in this case, is hours of sleep. Their amount of sleep can be measured without the pill ($C$) and with the pill ($T$).

One factor that might create a bias is the fact that people may have some psychological feeling after taking a medicine and that may have independent influence on their amount of sleep. This may or may not be related to the fact that they are being exposed more than once. In other words, such an effect can be present even if the two groups, $C$ and $T$, consist of different individuals.

Such a biasing effect can be taken care of easily by introducing a placebo in the experiment. In the initial observation, for $C$, the individuals may be given only a sugar-coated tablet (the placebo) and for $T$, the actual sleeping pill will be administered. An individual will not know whether it is receiving the actual pill or a placebo. In this case, even if there is a psychological effect, it is likely to be present in both $T$ and $C$. Hence, the effect will cancel out when a difference between the two observations is taken to obtain the treatment effect. If there is a reason to believe that it would be prudent to administer the two doses (placebo and the sleeping pill) in a random order so that not everyone receives the placebo first and then the pill, this can also be easily done provided the gap between the two observations is sufficient to nullify the presence of any multiple treatment effect (if it is true that a person taking a sleeping pill one day its effect could carry for few more days). This is known as 'single-blind experiment', signifying the fact that respondents are unaware (blind) about the administration of the treatment.

There may be a doubt about the experimenter's involvement in showing a positive effect. If this is so, it is possible to keep the experimenter also unaware of the actual administration of the treatment (the pill). Suppose, the sample size for the experiment is $n$. The experimenter will receive $n$ packets containing two similar looking tablets in each, one of which is a placebo and the other the pill whose effect is being measured. A third party will keep a track regarding the allocation to measure the effect. This is known as 'double-blind experiment'. The purpose behind both single-blind and double-blind experiments is to control bias.

**(ii)** Bias due to history effect and possible remedies

It is true that as the time gap between the two observations increases, the chance of occurrence of biasing effect due to history increases. If time gap between the two observations is sufficiently large and chance of occurrence of an unforeseen event is also high, this design will be vulnerable and an alternative design, discussed subsequently, should be considered. There is no specific remedy that can be suggested to reduce the effect of history that will be applicable in any before-after design. In some specific situations, a modification in the analysis might help.

Let us consider the example of evaluating the door-to-door campaign for institutional delivery. Suppose, the time gap considered in the experiment is 5 years. Also suppose that during the course of the programme, in the fifth year, the government announces a cash incentive programme for institutional delivery. This will act as an unforeseen event because it is likely to have its own influence on the output (per cent of delivery occurring in institution). Instead of analysing the increase in the per cent of institutional delivery after 5 years, which will have the biasing effect of the unforeseen event, it should be estimated based on information for the first 4 years, that is, leaving aside the birth that occurred during the last year after the event. Also, if additional information is collected, at the time of second observation, from women who have had a birth regarding the reason for choosing a particular place for delivery (home or institution), it can help in the process of reducing the bias.

**(iii)** Bias due to maturation, measurement and dropout effects and possible remedies

If the time gap between observations is large, the effects of maturation and dropout can also be substantial, if same sample is observed at both instances. Both maturity and dropouts tend to change the sample distribution in terms of certain disturbing variables at the time of the second observation. For example, maturation will certainly create a change in age distribution of the sample at the time of second observation. If age distribution is a disturbing variable, which is very likely, it will have a biasing effect. Dropout is also often related to age and hence influences the age distribution. This change in age distribution will also influence the overall average of the estimated parameter (at second point of time), which is used for obtaining the treatment effect.

The biasing effects can be avoided if the two samples (baseline and end line) are drawn independently. Even if the same sample is repeated, the biasing effect due to changes in the age distribution can be taken care of by applying the usual standardization technique (see Section 2.3.3).

It is necessary to take precaution regarding any changes in a measuring equipment. Any such change, as far as possible, should be avoided. A change in a measuring equipment (e.g., blood pressure or level of anaemia measuring equipment) that can introduce an unknown variation in the measurement should be avoided irrespective of whether a same sample is repeated twice or independent samples are drawn for a before-after study. Otherwise, there can be bias in the measurement of the estimated impact.

If a model is available that specifies the relationship between expected output and type of equipment, it can be used to make suitable adjustment to reduce the biasing effect. For example, if it is known (from other studies) that level of anaemia estimated by HemoCue will be higher by 3 mg/dl than that obtained by the BCC method, this adjustment can be incorporated while obtaining the impact of a treatment on anaemia. If average anaemia levels, in $C$ (estimated by BCC method) and $T$ (estimated by HemoCue), are, say, $x$ and $y$, respectively, the estimated impact ($I$) would be:

$$I = (y - x - 3)$$

#### 7.4.1.3 Estimation of standard error of estimated impact

The sampling variance of an estimated mean difference of a characteristic $y$ in $T(y_1)$ and $C(y_0)$ based on sample sizes $n_1$ and $n_2$, respectively (ignoring fpc) is

$$\frac{S^2}{n_1} + \frac{S^2}{n_2} = S^2\left(\frac{1}{n_1} + \frac{1}{n_2}\right) \tag{7.14}$$

where $S^2$ is the element variance of the population from where the two samples are drawn independently. The standard error of the estimated mean difference will be

$$\text{Standard error of } (y_1 - y_0) = S\sqrt{\left(\frac{1}{n_1} + \frac{1}{n_2}\right)} \tag{7.15}$$

If $n_1 = n_2 = n$, then

$$\text{Sampling variance of } (y_1 - y_0) = \frac{2S^2}{n}, \text{ and standard error is}$$

$$\text{Standard error of } (y_1 - y_0) = S\sqrt{\frac{2}{n}} \tag{7.16}$$

However, if same cohort is followed, the sampling variance will be

$$\text{Sampling variance of } (y_1 - y_0)^1 = \frac{2S^2}{n} - \frac{2rS^2}{n} = \frac{2(1-r)S^2}{n}$$

and

$$\text{Standard error of } (y_1 - y_0) = S\sqrt{\frac{2(1-r)}{n}} \tag{7.17}$$

The estimates of $S^2$ can be obtained from the sample (see equation 2.11). It can be seen from equation 7.17 that if a same cohort is followed, standard error of the estimate will be much less compared to that if samples of same sizes are taken independently (equation 7.16). This is because $r$, the correlation between $y_1$ and $y_0$, is likely to be positive and high. For example, if $y$ denotes income of a person, it is likely to be related at two points in time. If $r$ is 0.5, the standard error of $(y_1 - y_0)$ in the case of the cohort will be only 70 per cent of that if the two samples are drawn independently. However, it will also depend on the type of unit and the outcome used. If the units are clusters, as discussed in Chapters 2 and 3, the standard error of the estimate will increase because of the clustering effect, its value for estimating the treatment effect, in the case of independent samples, will be:

$$\text{Standard error of } (y_1 - y_0) = S\sqrt{\frac{2D}{n}} \tag{7.18}$$

where $D$ is the design effect as discussed in Chapter 3. For cohort, it will be

$$\text{Standard error of } (y_1 - y_0) = S\sqrt{\frac{2(1-r)D}{n}} \tag{7.19}$$

### 7.4.2 One Sample Each from *T* and *C* Observed at One Point of Time (Treatment-Control Design)

In this case, the observations on $T$ and $C$ are obtained simultaneously. Necessary details to obtain impact are discussed as follows.

#### 7.4.2.1 Description and estimation of effect

This is the second basic design that employs two distinct groups, one designated as $T$, where the treatment is administered and the other as $C$, the comparison where no treatment is administered. The outputs in both the groups are measured simultaneously and the treatment effect is obtained as

$$\text{Treatment effect} = y_1 - y_0$$

where $y_1$ and $y_0$ are as usual the estimated mean values of outputs in $T$ and $C$, respectively.

---

[1] It is well known that $V(y_1 - y_0) = V(y_1) + V(y_0) - 2Cov(y_1, y_0)$
Suppose, $n_1 = n_2$ and $s_0^2 = s_1^2 = s^2$
If $y_1$ and $y_0$ are independent then $V(y_1 - y_0) = \frac{2s^2}{n}$
If not, let the correlation between the two be 'r'.
Then, $Cov(y_1, y_0) = \frac{rs^2}{n}$ and $V(y_1 - y_0) = \frac{2s^2}{n}(1-r)$

#### 7.4.2.2 Biasing effects and remedies

Since the two groups are observed at the same point in time, the time effects are by and large absent. Although theoretically they are observed at same point, in practice, since collection of observation takes time, a gap may occur in some cases, particularly in experiments with large sample sizes. Similarly, there may be differential non-response between $T$ and $C$ that needs to be taken care of by suitable adjustment.

One major concern, in this design, is the presence of disturbing variables. In order to estimate true effect of the programme, the distribution of disturbing variables in $T$ and $C$ need to be similar. There are alternative procedures, including randomization to ensure this similarity and have been discussed separately in Chapter 8. If disturbing variables are known and can be controlled, the treatment-control design may be preferred to before-after design, particularly if time effects are to be avoided. However, the problem of considering external validity is more in this design compared to that in before-after design. Spreading a sample, for better representation of the population, may create difficulty in obtaining a balance between $T$ and $C$.

#### 7.4.2.3 Estimation of standard error of estimated impact

It is clear that the error or the sampling variance of the estimated treatment effect would be larger in this case, in comparison to that in the earlier design, where same group is observed twice. Since it refers to two distinct groups, the sampling variance for the treatment effect can be represented by

$$\text{Sampling variance} = 2\times\frac{S^2}{n}$$

$$\text{Estimated standard error} = s\sqrt{\frac{2}{n}}$$

where $s$ is an estimate of $S$. The standard error refers to that when individuals are the study unit, otherwise for clusters it will be

$$\text{Variance} = 2\times D\times\frac{S^2}{n}$$

$$\text{Estimated standard error} = s\sqrt{\frac{2D}{n}}$$

### 7.4.3 Two samples each from *T* and *C* observed at two points in time (before-after and treatment-control design)

Primary aim of this design is to make improvements in terms of reducing the biasing effects that are likely to be present in a before-after design. The gap in time present between the two observations, if not negligible, can introduce bias even if same or similar samples are considered for the design. The design is discussed below.

#### 7.4.3.1 Description and estimation of effect

The design is a combination of the two designs described earlier. It is also known as the 'before-after and treatment-control design' and tends to take care of the disadvantages of

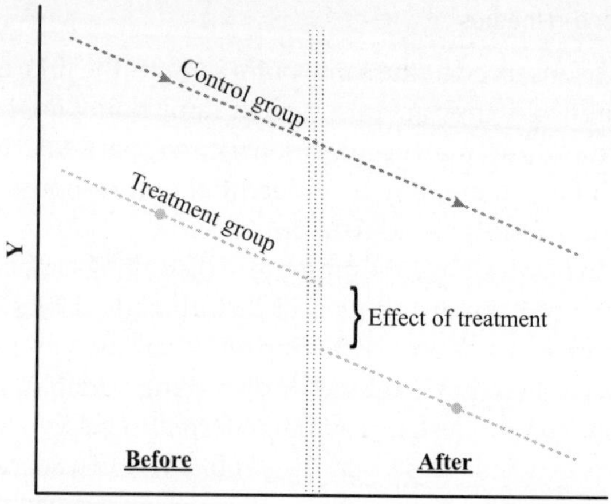


FIGURE 7.1 Effect of treatment in a 'before-after and treatment-control' design

the earlier designs. In this case, two groups *T* and *C* are selected and initial observations are made before introducing the intervention. Let the two outputs be denoted as $y_{T0}$ and $y_{C0}$. The treatment or intervention is then introduced in *T* for a specified period of time. After this, again observations on *T* and *C* are made and let they be denoted as $y_{T1}$ and $y_{C1}$. The design can be represented graphically as in Figure 7.1.

Now, as it is clear from Figure 7.1, the additional change in the treatment group *T* compared to the control group *C* is due to the treatment.

We first compute the differences between *T* and *C* at each of the two points in time denoted by $d_0 = (y_{T0} - y_{C0})$ and $d_1 = (y_{T1} - y_{C1})$. Since *T* and *C* are selected in such a way that most disturbing factors are similar in both areas, at the time of initial observation any difference in outcome between *T* and *C* is possibly due to unobservable factors. At the time of final observation, if there is a positive effect of the intervention on the outcome of interest, difference between *T* and *C* will largely be due to the intervention as well as due to unobservable factors. By subtracting $d_0$ from $d_1$, we will be able to separate out the effect of the unobserved factors and estimate the effect of intervention. Hence, in this case the treatment effect can be obtained as the difference between these two differences, that is,

$$\text{Treatment effect} = d_1 - d_0 = (y_{T1} - y_{C1}) - (y_{T0} - y_{C0})$$

An essential assumption for obtaining true effect of the intervention using this method is that any change other than the presence or absence of the intervention in the two groups will be same in both the groups. The difference between two differences ($d_0$ and $d_1$) is purely due to the intervention. In other words, it assumes that there is no interaction between the components of time effect, such as history, pre-test and maturation, measurement change and dropout effects, and the effect is of the treatment. That is, the individual effects of different components of time would be similar in *T* and *C*. This method of estimating treatment effect is popularly known as difference-in-differences (DiD) method.

#### 7.4.3.2 Biasing effects and remedies

We can write the formula for estimating treatment effects by simply rearranging the terms as

$$\text{Treatment effect} = (y_{T1} - y_{T0}) - (y_{C1} - y_{C0})$$

It is clear that $(y_{T1} - y_{T0})$, which in fact is the estimate of effect in the before-after design, gives the treatment effect only under the assumption that the main effects of history, pre-test, maturation, measurement change and dropouts are absent. By including another difference $(y_{C1} - y_{C0})$, this 'before-after and treatment-control design' avoids that assumption by removing biasing effects of the factors, such as history, pre-test, etc. For this reason, before-after and treatment-control design has been used widely and regarded as one of the most powerful and ideal design in social sciences.

Let us now discuss the assumptions in greater details. Consider the history effect, the design assumes that it should be similar in *T* and *C*. It is clear that the history effect depends on the time gap between the two observations. That is, the time gap between $y_{T1}$ and $y_{T0}$ and between $y_{C1}$ and $y_{C0}$ should be same. While this is implied, it needs to be implemented as such. For example, if the time taken to complete the data collection for any group takes 1 month and the data collection is done in the following order: first in *T* (i.e., $y_{T0}$ is estimated), then in *C* ($y_{C0}$), again in *C* ($y_{C1}$) and then in *T* ($y_{T1}$). The time gap between $y_{T1}$ and $y_{T0}$ will clearly be higher by 2 months, compared to that in $y_{C1}$ and $y_{C0}$. This can make a difference in the history effect in *T* and *C* and hence can create a biasing effect. Even if the data collection begins simultaneously in *T* and *C* both in the initial and the final rounds, their completion time may differ due to some reasons. Such differences need to be minimized.

Even if the time gap between two observations is the same between *T* and *C*, the differences in magnitude and nature of the dropout rates in *T* and *C* can create a biasing effect. The techniques of incorporating the non-response error through weights (mentioned in Chapters 2 and 3) should be employed in both the groups to avoid such biasing effects in the analysis. As far as possible, a similar technique should be used for both.

If the time gap in data collection is same between *T* and *C*, the maturation effect should be the same in both $(y_{T1} - y_{T0})$ and $(y_{C1} - y_{C0})$. This is also true in the case of change in measurement technique. If same equipment is used for *T* and *C* during the initial and final rounds of data collection, the differences observed in $(y_{T1} - y_{T0})$ and $(y_{C1} - y_{C0})$ would be the same and it would be cancelling out while obtaining the treatment effect.

**(i)** Bias due to contamination

There can be a completely different type of bias in this design, known as the 'spillover effect' or 'contamination effect'. It occurs if there is a mixing between the units in *T* and *C* so that the benefits, if any, of the treatment percolate in the control group; it can influence the behaviour of units in the control group. For example, while evaluating the effect of door-to-door campaign on institutional delivery, if units belonging to *C* come to know about the campaign (treatment) from someone in *T*, it can change their outlook about the need to have institutional delivery. This can unduly increase the per cent of institutional delivery in $C$ and hence negatively bias the treatment effect. Larger the time gap and more the geographical proximity between the units of $T$ and $C$, more is the chance of the spillover effect on treatment effect. This biasing effect is absent in a before-after design but may also be expected in a treatment-control design described before (there can be a time gap, though generally small, between $T$ and $C$ and depending upon how quickly the contamination can influence a output, the spillover effect may be present).

To reduce this effect, it is suggested to keep a fair geographic distance between the units in $T$ and $C$. The selection of units needs to be done carefully because it is also desirable that $T$ and $C$ should be as similar as possible, in fact, it is better if they are random samples from the same population. However, keeping a large geographic distance between $T$ and $C$ can be problematic as such arrangements are likely to need more resources than usual.

Possibility of reducing history effect through collection of additional data has been mentioned, in the case of a before-after design. Similar possibility might also exist to alleviate the spillover effect, in this design. Considering the same example, collection of additional data on reasons for preferring institutional delivery might help in assessing the spillover effect.

**(ii)** Bias due to interaction between treatment and pre-test effects
One effect that can be a cause of worry, particularly true in psychological experiments, is about the fact that the pre-test effect need not be same in $T$ $(y_{T_1} - y_{T_0})$ and $C$ $(y_{C_1} - y_{C_0})$. In other words, there can be an interaction between the treatment and the pre-test effects. Although the design controls the main effects of factors arising due to the time gap, it does not control the interactions. Generally, interaction effects are negligible but it may not be so in some situations where units' familiarity with treatment can make substantial changes in $T$ than in $C$. To alleviate this, a design that combines this design and post-test only design[2] has been suggested (Campbell and Stanley, 1963; Namboodiri, 1970; Solomon, 1949).

#### 7.4.3.3 Estimation of standard error of estimated impact

The sampling variance in the present design will be larger compared to the two designs discussed earlier since we will have to compare four means (or proportions). It can be approximated as follows:

$$\text{Variance} = 4 \times (1-r) \times \frac{S^2}{n}$$

$$\text{Estimated standard error} = 2s\sqrt{\frac{(1-r)}{n}}$$

when individuals are the study unit, and if clusters have been used, sampling variance can be expressed as

$$\text{Variance} = 4 \times (1-r) \times D \times \frac{S^2}{n}$$

[2] A post-test design, unlike the before after design, considers observation on the treatment ($T$) only.

$$\text{Estimated standard error} = 2s\sqrt{\frac{(1-r)D}{n}}$$

## 7.5 OUTPUT AND ITS TIMING

It is necessary in any evaluation design to decide about its output or the dependent variable on which the effect of a treatment would be measured. Along with this, it is also required to define the time when an output is to be measured. That is, to decide the time when a post-test observation or end-line survey needs to be done in a pre-post- or before-after design. The two, that is, output and the time when it is to be measured to get the change, are related to a large extent. Let us consider a design that requires evaluation of an advertisement to popularize a particular brand of shampoo. The output could be per cent of people who preferred the variety. In this case the time of measurement of the output can be as soon as possible, even on the same day. But if the output considered is the volume of sale of the product, the time of measurement has to be delayed somewhat maybe within the next 30 days. In a family welfare programme, suppose a study is to be designed to understand the effect of door-to-door provision of temporary methods of contraception. If the output on which the programme is to be evaluated relates to acceptance of temporary method of contraception, then the time of measurement need not be large. But if it is to examine the duration of use of a method or impact on fertility, the gap should be larger.

Following two examples will help in understanding the importance of timing or how a decision on time can influence the impact. In Sweden, in 1967 a study was conducted to assess the impact of changing the traffic rule for driving from left side to right side of roads. The accident rate was surprisingly found to have dropped when an observation was taken after 1 month (after introduction of the new rule). But when the observation period was extended, it was found to increase. The point is, initially people exercised caution while driving because of the change in the rule and it resulted in the decline. But once they were habituated and the extra caution disappeared, the real effect started to emerge (as described in Kish, 1987).

In Romania, in 1960s the abortions were considerably restricted by the law. As expected, initially the fertility shot up after a year of introduction of the new law. But it gradually started to decline afterwards, largely because the incidence of illegal abortions began to increase. Both these examples highlight that there can be an initial reaction to an intervention programme that may vitiate its true effect and it would be important to allow time so that the reaction wares off (Kish, 1987).

## 7.6 KEY POINTS

This chapter focuses on the formation of contrasts to develop an evaluation design.

- The basic approach in measuring the effect of a treatment is to compare observations on a given output on two groups of study units, one with the treatment ($T$) and the other the control (C), that is, without the treatment.

- It is required that the outputs observed for $T$ and $C$ are similar in the absence of the treatment.
- The variables that could disturb the similarity in observations between $T$ and $C$ in the absence of the treatment and hence can introduce bias are categorized into: a) 'disturbing variable' and b) variables due to a difference in 'time' in observing the two groups.
- Disturbing variables are those that, if differently distributed among $T$ and $C$, tend to produce a difference in the two outputs. A difference in the timing of observations or the time effect occurs due to repeated exposure of a unit (pre-test effect), abrupt changes in certain events (history effect) and maturation, changes in measurement, dropouts, use of multiple treatment and regression effects.
- In an ideal design, the comparison is made using one group observed at the same point of time. Since there is one group, the selection bias does not exist. So are the time effects, as both observations are made simultaneously. Alas, opportunities are rare to have such a comparison.
- In a before-after design the similarity, in the absence of a treatment, is attempted by comparing either the same group of study units at two points of time or having the two groups drawn randomly from the same population at two points of time.
- A before-after design, therefore, tends to be strong in controlling the disturbing variables but weak in tackling the time effect.
- A treatment-control design, on the other hand, considers two distinct groups for the comparison but observes them at the same point of time. The similarity between the two observations, in the absence of treatment, is tried either by having the two study units drawn randomly from the same population at the same time or by matching the units.
- Treatment-control design, therefore, tend to be strong in controlling the time effects but weak in ensuring similarity in the disturbing variables.
- There is another design that is a combination of the above two. In this design, observations on two groups $T$ and $C$ are made at two different points of time. It is called before-after and treatment-control design.
- This design, regarded as an ideal one in social science sector, has the potentials to take care of the main effects of all disturbing variables as well as the time effects and hence controls all the biasing effects, except for the presence of interaction effects, if any.

Chapter 8

# Designs for Causal Effects: Allocation of Study Units

## 8.1 INTRODUCTION

Allocation of units to treatments is crucial in designing a study on impact evaluation. The strength of an experiment depends, in addition to formation of the comparison groups (discussed in the previous chapter), on how units are allocated to the two groups. Together they make a design complete in the assessment of the causal effect.

The basic objective in a design is to create a control and a treatment group in such a way that sans the treatment the two would be similar and produce outputs that would be statistically equal to each other. The similarity between the two groups can be achieved in a few different ways. This chapter describes the different types of allocation along with their advantages and disadvantages.

In this context, another approach known as propensity score matching has been gaining popularity. This method, popularly known as Rubin Causal Model, which assists in strengthening a design, has also been discussed. The issue of two distinct situations, depending on whether individuals or clusters are the units of allocation, has been given adequate attention.

## 8.2 ALTERNATIVE TOOLS TO ATTAIN BALANCE

To achieve a balance, one needs to ensure that both treatment and control groups have similar distribution in terms of all the disturbing variables (covariates). There are two distinct approaches to balance the disturbing variables. They are: randomization and matching. The matching can be performed in two ways: a) group matching or stratification and b) pair matching.

### 8.2.1 Randomization

The objective can be achieved by ensuring that both treatment and control groups are randomly drawn samples from a population under study. That is, either they are drawn with an equal probability selection method EPSEM or, if not, individuals are weighted appropriately so that each provides an unbiased estimate of the population parameter. There is a need to emphasize this, since the issue of unbiasedness of an estimate can be overlooked, particularly when non-response makes the two samples to deviate from an EPSEM. If the pattern of non-response is different in the two samples (treatment and control) it can introduce bias in the estimation of the impact. The difference in the estimates obtained from the two samples will not show the treatment impact, unless the samples are weighted properly to remove the biasing effect created by the non-response.

For example, in a before-after design, the observations made at a later point in time are more likely to suffer from the non-response and hence would require the application of weights. The random selection facilitates making statistical inference about the impact. Although the details of such analyses have been discussed in the next chapter, it is pertinent to have a brief idea about it. Let $\bar{x}_T$ and $\bar{x}_C$ denote the sample means of the treatment and the control groups, respectively, with $n_T$ and $n_C$ as the respective sample sizes. Since they are drawn from the same population, the expected value of their difference will be zero, if the treatment has no effect. In other words, $\bar{x}_T - \bar{x}_C$ will provide the estimated value of the treatment effect. However, since both are samples, it is necessary to provide leeway for the sampling fluctuations. That is, to check whether the difference is large enough to be considered as to have not occurred due to the sampling fluctuations.

The test statistic, $t$, defined as

$$t = \frac{\bar{x}_T - \bar{x}_C}{S\sqrt{\left(\frac{1}{n_T} + \frac{1}{n_C}\right)}} \tag{8.1}$$

can be used to understand the statistical significance of the difference, that is, whether the difference is real and is the impact or it is because of the error due to sampling. $S$ is the square root of the element variance in the population, which, as mentioned in Chapter 2, can be estimated from the sample (denoted by $s$). The denominator of equation 8.1 indicates the standard error of the estimate of the numerator. That is, we should allow this much difference (error) because of considering the sample values. It is obvious that smaller the value of the denominator (standard error of the estimate), greater is the chance that the difference observed in the numerator would be real and due to the treatment rather than occurring due to sampling fluctuations, which means greater is the precision of the estimated impact or less is the Type I error (Section 7.3).

Another worth noting point is that larger the sample size, which is related to the degrees of freedom ($df$), to estimate the difference, for a given $S$, smaller would be the standard error of the estimate. The value of $df$ in this case will be

$$\begin{aligned} df &= (n_T - 1) + (n_C - 1) \\ &= (n - 2) \end{aligned}$$

where $n = n_T + n_C$ is the total sample size. It may be recalled from Chapter 2 that the estimate of $S$ would be based on $(n_T - 1)$ individuals selected in the case of the treatment and $(n_C - 1)$ for the control group. In other words, larger the sample size higher would be the precision of the estimated impact. Generally, sample size for $T$ and $C$ are kept same. However, the differential non-response or dropouts are likely to make them unequal.

### 8.2.2 Stratification

When element variance ($S^2$) is large in a population, it would tend to make the design with randomization less precise. This is because a large standard error means a relatively larger difference between $\bar{x}_T$ and $\bar{x}_C$, which would be regarded as due to sampling fluctuations (refer equation 8.1) and not a true difference. A stratified design would help in improving the precision. Since the variables for stratification should be related to the response variable, they generally happen to be the disturbing variables.

Let us consider a study on measuring the effect of a food item on weight gain. Age of a person is likely to be a disturbing variable. A random selection from the population will consist of individuals of different ages and hence create a larger variability in the weight gain. As mentioned in stratified sampling, stratification by age would reduce the variability. The population will be divided into different strata, and samples from each of the strata will be drawn and allocated randomly to the treatment and the control group. The impact, the difference in the response variable between $T$ and $C$ will be obtained separately for each stratum. Within each stratum, age is automatically held constant (matched) which has an intuitive appeal. Let us assume that age is considered in two categories. Also, let $\bar{x}_{1T}$ and $\bar{x}_{1C}$ be the mean of the response variable in $T$ and $C$ with sample sizes as $n_{1T}$ and $n_{1C}$ in stratum 1. Using equation 8.1, it is possible to check the impact of the treatment in stratum 1 by calculating the statistic $t$ as

$$t = \frac{\bar{x}_{1T} - \bar{x}_{1C}}{S_1\sqrt{\left(\frac{1}{n_{1T}} + \frac{1}{n_{1C}}\right)}} \tag{8.2}$$

where $S_1$ denotes the element variance in stratum 1. Its *df* will be $(n_{1T} + n_{1C} - 2)$. Similarly, the impact can be obtained for the other stratum with the corresponding estimates as $\bar{x}_{2T}$, $\bar{x}_{2C}$ and $S_2$ and sample sizes as $n_{2T}$ and $n_{2C}$. The procedure remains the same for more than two strata. To get the impact at the overall level, one needs to obtain combined mean response separately for the treatment as well as the control. An average of $\bar{x}_{1T}$ and $\bar{x}_{2T}$, for example, can provide an idea about the overall mean response for the treatment in the sample and similarly for the control. However, for obtaining an estimate of the impact for the population, the stratum estimates are to be weighted according to their proportion in the population (as discussed in Section 2.3.2.1). The value of *df* to test the statistic would be

$$\begin{aligned} df &= [(n_{1T} + n_{1C} - 2) + (n_{2T} + n_{2C} - 2)] \\ &= (n - 4) \end{aligned}$$

where $n$ denotes the total sample size $(n_{1T} + n_{1C} + n_{2T} + n_{2C})$.

The guidelines for considering variables and their categories for the stratification have been outlined earlier while discussing the stratified sampling (see also Wells, 1975). The suggestion that it is better to consider more variables (disturbing variables) with less categories for each than including few variables with larger number of categories for each holds good here also.

If number of units for an experiment is small and number of variables that needs to be controlled is relatively large, one faces a problem in stratification or matching. This can happen especially when clusters are the units of selection. Since the disturbing variables are often related to each other, it is possible to obtain a few independent factors using the principal component analysis. The few components so obtained can then be subdivided to form the required number of categories that can be used for the stratification (Hagood and Bernert, 1945).

Although such an approach can be good for exercising control, it has one disadvantage. Since the disturbing variables are combined suitably to get the components, identity of the individual variables get diluted. Therefore, the effects, particularly the interactions, if present, cannot be interpreted properly. At best it can be mentioned that an impact is likely to vary according to the levels of the disturbing variables. But it may be difficult to identify the particular disturbing variable (risk factor) responsible for such a variation.

### 8.2.3 Pair Matching

This happens to be a popular method for designing an experiment. Conceptually, it is similar to stratification. The groupings are done extensively so that two units are selected from each stratum. The pairs so formed are regarded to have characteristics (all characteristics or disturbing variables perceived to be important while studying the impact) similarly distributed between them. Sometimes such pairs occur naturally. One case of such natural pairing is mentioned in an example on evaluating durability of a shoe sole in the previous chapter. Two legs of a person become a perfect pair that tends to control almost every conceivable disturbing variable. In many other situations, twins can form a very viable option for the consideration of pairs.

Let $k$ such pairs be made from $n$ (= $2k$) available units. Also, $x_{iT}$ and $x_{iC}$ are the response variables for treatment and control in $i$th such pair and

$$d_i = x_{iT} - x_{iC}, \text{ for } i = 1, 2, \ldots, k \tag{8.3}$$

We, therefore, have $k$ observations on the difference, each of which measures the impact of the treatment. It is possible to test whether $d_i$ is equal to zero (for details see Chapter 9). The value of $df$, in this case, will be $k - 1$.

## 8.3 ADVANTAGES AND DISADVANTAGES OF THREE TOOLS

A natural question that arises in a design is to choose a particular tool for the allocation of units to $T$ and C. It is therefore pertinent to understand the advantages and disadvantages of different available tools.

### 8.3.1 Randomization

Randomization guards against the biasing effects of all known as well as unknown variables. Experiments with random assignment of units to *T* and *C* are regarded as the true experiments.

To better understand the advantage of randomization, let us consider the example of finding the effect of a sleeping pill. Suppose, random allocation of units to those who agree to have the pills (*T*) and those without the pill (*C*) is not possible and the two groups are matched, that is, those who agree on their own to take the pill will be chosen to form *T* and *C* will be the matched group. Although most of the disturbing variables could be controlled, it will be difficult to control one variable that might be very important. It may so happen that most of those who agree on their own volition to take the pill were having a sleep disorder and therefore felt the urge to join the experiment. Not only that it will affect the internal validity, the external validity would also be jeopardized.

If randomization was used in the allocation this would not be so. All those with such disorder will not be en masse in the treatment but will get distributed to both *T* and *C*. It can be noticed that if one used the 'before-after and treatment-control' design, the initial difference between *T* and *C* ($d_0$) is likely to rectify the bias when subtracted from the difference between *T* and *C* at a later period ($d_1$).

There is one difficulty in randomization that the sampling variance of an estimate could be large, particularly if a population is heterogeneous. Hence to detect an impact (see equation 8.1), large sample size will be required, which will increase the cost. Also, there are practical issues, mentioned earlier, that restrict the use of randomization.

### 8.3.2 Matching

Matching is the other potent tool to provide balance between *T* and *C*, in the absence of randomization. As mentioned earlier, both stratification and pair matching involves matching. The only difference between the two is that each pair of units are matched in terms of all the disturbing variables before allocation to *T* and *C* in the latter, whereas a number of units belonging to a specific group (strata) are used for the allocation in the former.

#### 8.3.2.1 Stratification

Apart from reducing the effect of heterogeneity, stratification facilitates obtaining idea about the interaction between a disturbing variable and a treatment. To illustrate the issue of interaction, let us consider an example (for details, see McKinlay, 1977). In a study, the interest is on measuring the effect of an antitoxin in influencing the survival status of patients suffering from a disease. A risk factor (disturbing variable) considered is the condition of a patient (with severe or moderate symptom) at which the antitoxin is administered. To simplify, the following variables are defined:

$x_{1T}$ = proportion of patients survived among those who were administered antitoxin (treatment) with severe symptom,

$x_{2T}$ = proportion of patients survived among those who were administered antitoxin (treatment) with moderate symptom,

$x_{1C}$ = proportion of patients survived among those in severe symptom but were not given antitoxin (control),

$x_{2C}$ = proportion of patients survived among those in moderate symptom but were not given antitoxin (control).

The formation of two strata were done according to the condition at which patients were brought when the decision on administering antitoxin or not was taken. Clearly, the effect of antitoxin can be computed separately for the two strata.

Effect of antitoxin when administered at severe condition

$$= (x_{1T} - x_{1C})$$

Effect of antitoxin when administered at moderate condition

$$= (x_{2T} - x_{2C})$$

If the two effects are different (statistically), it is said that there is an interaction between the treatment and the symptom in producing the impact.

Finding an interaction is itself important. Suppose, it is found that the antitoxin is effective only among those who were being administered in a moderate symptom $(x_{2T} - x_{2C})$. It is not so among those who were administered in a severe condition. Suppose, the individual effects are not computed and only the overall effect is analysed. The overall effect computed on the basis of appropriate weights for the two strata may or may not be statistically significant. Depending on the result, one would interpret whether the antitoxin (treatment) has an effect or not. In either case, there will be a loss of information. Even if the overall effect is significant, the point that it is important to administer the antitoxin at moderate condition will be missed.

For many programme evaluation or marketing research studies, the objective may be to understand the effect of an incentive programme (monetary or non-monetary) on increasing the acceptability or utilization of a product. For example, exposure to benefits of institutional delivery on its acceptability. In such cases, generally the socio-economic characteristics are not only important disturbing variables (affecting the acceptability) but also are likely to interact with the treatment (exposure to the programme). It may be more suitable to use stratification in such situations. This can help in focusing attention on identifying specific groups of individuals who would need to be targeted.

As discussed in Chapter 2, the sampling variance of an estimate is likely to be less in stratified sampling compared to the SRS. Hence, stratification if carried out effectively can serve the dual purpose of increasing the precision of an experiment as well as provide idea about the presence of interaction.

#### 8.3.2.2 Pair matching

Intuitively, it appears that the precision of pair matching will be high. The variance of the difference is likely to be small. Let us consider the example of a shoe sole (discussed in Section 7.2). Since the same person is wearing a shoe with treatment (sole that is being evaluated) in one leg and shoe used as control (sole generally used) in the other, $d_i$ will show the impact. The impact is likely to be similar, irrespective of the socio-economic and environmental conditions (considered to be the disturbing variables). It is true that the disturbing variables will influence the value of the response variable, but it will be similar for both $x_{iT}$ and $x_{iC}$. For example, the duration of use is expected to be less for a person who walks for more hours compared to one who moves generally by car. But it is expected to affect both $x_{iT}$ and $x_{iC}$ similarly so that $d_i$ is likely to be independent of the hours of walk.

In other words, the variance of $d_i$ is likely to be less compared to that if the allocation is done either completely at random or by stratification.

Even though the available *df* is much less in pair matching, the standard error of the estimate is likely to be lower compared to the other two methods (randomization or stratification). The precision of a pair matching depends on the effectiveness of matching. It has been mentioned that matching is effective when disturbing variables considered for matching show strong correlation with the outputs of a study (Cochran, 1950; Worcestor, 1964; Youkeles, 1963).

It is clear that the variance of $d_i$ is likely to be less because the values of $d_i$ are likely to be similar even if values of a disturbing variable change. But this will not be so if there is an interaction between a disturbing variable and a treatment. Hence, the presence of interaction will tend to increase the sampling variance in pair matching. The presence of interaction not only influences the precision but also can result in an important disadvantage in the statistical inference, if pair matching is not used carefully (McKinlay, 1977).

The aspect of cost for conducting an experiment can also be important in pair matching. The cost of creating matched pairs can be substantial in some cases. Searching for matched pairs can be laborious and can lead to wastages. Many units may have to be discarded because of the absence of suitable matched units. The magnitude of wastage depends on the criteria used for matching.

If age is a factor for matching, the search can be long and leading to higher wastage when two persons of same age are to be matched as against matching two persons belonging to same age group such as 15–19, 20–24, etc.

It is also true that if any unit in a pair gets dropped due to some reasons, the other unit in the pair also becomes redundant. This will not be the case in the other two approaches. Such non-response or dropouts, if not large, may not create problems in cases where individuals are the elements in a study. But it may result in a big loss if clusters are used. In a cluster randomized trial matching is done at cluster level. Since not many clusters are included in a study, loss of two clusters can have significance. It may not be too uncommon to find a whole cluster opting to drop out from a study.

Generally, wastage in matching or dropouts is not mentioned in a study. However, as mentioned by McKinlay (1977), one needs to have about 25 per cent higher sample size for accommodating the loss in conducting a pair matching. This type of loss can also mean compromising the generalizability of a study. In other words, pair matching may be strong in internal validity but weak in external validity.

## 8.4 CHOICE OF STUDY UNITS

Study units can be of two types: an individual element or a cluster, that is, a group of individual elements. The choice of unit is however not random; it depends largely on objectives of a study. For example, to find the effect of smoking on lung disease, the focus should be on individuals who smoke vis-a-vis who do not smoke. However, if the objective is to obtain the effect of a ban on smoking in public places on prevalence of smoking, the focus should be more on assessment at the community level.

On many occasions, both elements and clusters can be considered as a study unit. Let us consider the case of finding the effect of a door-to-door campaign on benefits of having births at an institution on prevalence of institutional delivery (the objective can also be to understand its influence on maternal and child health). It is possible to consider women, specifically of reproductive ages, and form two groups so that one will be given the treatment (exposed to the campaign) and the other without it. However, selection of a representative sample and organization of the intervention would require larger resources if individual women rather than clusters of women are considered as the units. In both the cases, the output remains the prevalence of institutional delivery. Hence, apart from the objectives, the study of feasibility plays a major role in choice of study unit.

Use of cluster has also certain other advantages. First, cluster sample is easier to draw compared to selection of individual elements. Often implementation of an intervention programme is convenient when administered at a cluster level rather than at an individual level. A cluster as a study unit can also provide better protection from the spillover effects (contamination) compared to that when individuals are the units because chances of contact between two individuals living in the same cluster is higher than when they live in different clusters. Moreover, if individuals are allocated randomly to treatment and control groups, chances are there that two individuals in two different treatment arms will belong to a same cluster. As already mentioned, ethical issues often arise when some individuals are randomly exposed to a treatment and some are not exposed. For example, in the evaluation of the campaign for institutional delivery, an ethical issue might be that everyone should be exposed to this beneficial information. At the cluster level, such issues tend to become less. Also, a 'stepped wedge' design that ensures that the intervention occurs gradually in every cluster can further help in addressing the issue (Hayes and Moulton, 2009).

There are, however, disadvantages of using clusters as units. Because of the effect of clustering (Chapters 2 and 3), the sampling variance of an estimate increases which tends to reduce the effective sample size. In other words, sample size has to be increased to get a similar precision that would be obtained had individuals been selected directly from a list. It is also difficult to exercise control when clusters are the units. For example, education is easy to match exactly between two individuals but difficult at a macro level. The mean level of education can only be matched, that too if prior information is available.

### 8.4.1 Procedure of Allocation of Units/Clusters

This section discusses the actual selection procedure under the three methods, that is, randomization, stratification and pair matching.

#### 8.4.1.1 Randomization

Once the sample sizes for the two treatment arms ($n_T$ and $n_C$), in an experiment are decided, it is necessary to allocate the units randomly to the two arms. Let us consider that $n_T = n_C = n$.

The total number of units to be selected is $2n$ from a population consisting of $N$ units and they are to be distributed randomly to $T$ and $C$ so that they have equal number of units. When unit refers to individuals and $n$ is also large (greater than say 30), the SRS approach (discussed in Chapter 2), such as selection with the help of random numbers or systematic sampling can be employed.

For illustration, let us consider that $N$ is 6000, and 30 units each are to be allocated to $T$ and $C$. One can consider the interval $I$ as

$$I = \frac{6000}{30} = 200$$

that is, two units are to be selected from 30 groups consisting of 200 units each. Two random numbers between 1 and 200 (say, 29 and 163) can be selected to get the samples. That is, $T$ and $C$ will consist of samples corresponding to serial numbers (29, 229, 429, …, 5829) and (163, 363, ..., 5963). If units in the frame are arranged geographically, which is generally the case, the sample will give an adequate geographic spread. In addition, the spread will be similar for both $T$ and $C$. In fact, if geographic location is important, it might be worthwhile to form 30 pairs as (29 and 163), (229 and 363),…, (5829 and 5963).

In the case of programme evaluation, more often the situation calls for selecting $n$ clusters each from available $2n$ clusters for the allocation. Usually, $n$ will be small and a random selection as above may not suffice. The matching needs to be done at the cluster level and the available information are generally socio-economic variables and geographic area. Variables to be matched may be quite a few. For rural areas in India, one may consider the distribution of villages by literacy, caste composition, etc. For each matching variable, one can decide a caliper. Let us consider the case of matching villages according to level of literacy. With limited number of villages for an experiment, it may be difficult to find two villages having exactly identical level of literacy. Instead, one may specify a range (such as 10 per cent on either side of an average level) so that average literacy of two villages falling within the range (as between 9 and 11 or 72 and 88) will be regarded as having been matched in terms of literacy.

If $n$ is small, as it is true in cases where clusters are the units of a study, complete randomization may not serve the purpose of providing adequate matching between two treatment groups ($T$ and $C$). Some additional restrictions on the allocation can facilitate this. This is known as restricted randomization (for a detailed discussion, see Hayes and Moulton, 2009).

#### *8.4.1.1.1 Restricted randomization*

As has been mentioned in Chapter 2, SRS gives equal chance of selection to each possible sample. If $2n$ villages are to be divided equally between $T$ and $C$, using SRS, the total possible samples of size $n$ would be given by

$$\frac{(2n)!}{n!n!}$$

Since they are equally likely to occur, chances of allotting any one to $T$ and the other to $C$ would be equal. However, as indicated in Example 1.1, there are disadvantages in adopting a SRS, particularly if $n$ is small. Possibilities exist that the samples allotted to $T$ and $C$ will differ considerably in terms of the variables to match.

Also, assume that there are two variables, namely, per cent of literate population and village size that require matching. The two variables for the four villages $v_1$, $v_2$, $v_3$ and $v_4$ are (20, 200), (65, 500), (70, 550) and (22, 180), respectively. That is, village $v_1$ consists of 200 persons and 20 per cent of them are literate, and so on in the other three villages. It can be seen that villages $v_1$ and $v_4$ are quite similar to each other in terms of the two disturbing variables, whereas the same is true for $v_2$ and $v_3$. Now, if the selection is done using SRS, there will be

$$\frac{4!}{2!2!} = 6$$

possible samples of size 2, which are $v_1v_2$, $v_1v_3$, $v_1v_4$, $v_2v_3$, $v_2v_4$ and $v_3v_4$. It may be alright if the pairs $v_1v_2$ and $v_3v_4$ are obtained for the allocation to $T$ and $C$. Similar is the case with the pairs $v_1v_3$ and $v_2v_4$. But if $v_1v_4$ and $v_2v_3$ are obtained for the allocation, matching will not be exercised adequately. Whichever treatment gets $v_2v_3$ will have a favourable situation in terms of having a larger proportion of literates staying in bigger villages (bigger size villages are more likely to have better facilities compared to those that are of smaller size).

In restricted randomization, occurrence of $v_1v_4$ and $v_2v_3$ will be given a zero chance or they will be excluded from the allocation. In other words, out of the four remaining combinations, $v_1v_2$, $v_1v_3$, $v_2v_4$ and $v_3v_4$, one will be selected giving equal chance to each four and allocated to, say, $T$. The remaining two villages will be for $C$.

Although the procedure of restricted randomization is apparently simple, its actual implementation is not so. Even if, say, $n$ is 12, there will be 924 possible selections of six clusters each for $T$ and $C$. It will require examining the restrictions for all these 924 combinations. As $n$ increases, the possible combinations increase rapidly. The help of a computer programme is required to implement the procedure.

Imposing restrictions on the allocation needs to be done objectively. It may be mentioned in this context that in stratified sampling too some restrictions on selection is enforced. It is prudent to remain vigilant, while finding the eligible number of selections in a restricted randomization, so that validity of an experiment is not questioned. It is necessary to check the following two conditions (Hayes and Moulton, 2009):

**(i)** If original number of possible allocations is large, a minimum of 100 allocations should be available after imposition of restrictions.

**(ii)** The chance of selection of a unit (cluster) from an eligible list of selection should remain same for both $T$ and $C$.

Suppose, there are $K$ combinations finally available after the restrictions have been imposed. Let $m_{iT}$ and $m_{iC}$ denote the number of times $i$th unit occurs among $T$ and $C$, respectively. Then $m_{iT}$ should be equal to $m_{iC}$, that is, every unit should have equal chance of being included in either $T$ or $C$. If violated, this can introduce bias to the design. It can be checked in the above example out of the four available combinations, after excluding $v_1v_4$ and $v_2v_3$, the probability of including a village $v_i$ to either $T$ or $C$ is 1/2.

To have further checks on validity, apart from the occurrences of a single unit, one can also examine the occurrences of pair of units and ensure that they should also not vary much between the two treatment groups.

In the case of quasi-experimental designs, units or clusters belonging to $T$ are already defined. One can adopt randomization to select C. The propensity score matching discussed subsequently (Section 8.5.1) can be employed in such a situation.

#### 8.4.1.2 Stratification

Allocation for an experiment would require formation of strata based on desired variables and their categories. The numbers of units $n_{iT}$ and $n_{iC}$ in $i$th strata belonging to treatment and control, which are generally considered equal, are selected by randomization or restricted randomization.

In the case of an evaluation design with matching (either at unit or cluster level), where, in general, units in $T$ are self-selected, they are subdivided into the different strata. That is, number of treatment units in the $i$th strata, $n_{iT}$, is known. The $n_{iC}$ units are selected from the available pool of the units, if possible, using randomization.

#### 8.4.1.3 Pair matching

In this case, two units one for $T$ and the other for C comprising a pair are generally selected randomly from a population. Ideally, if $N_i$ units are available in a population that are already matched in terms of all the disturbing variables, two units can be selected with randomization and allotted to $T$ and C randomly.

In the case of evaluation designs with matching, if for a unit in $T$, there are more than one unit in C available that can match with its characteristics; one of them can be selected randomly. The 1:1 matching is not mandatory, and it is possible to have 1:2 or even 1:$m$ matching. In other words, for each treatment there can be $m$ matched units in C, particularly if cost of inclusion of one unit in C is less compared to that of $T$ (Breslow and Day, 1980).

## 8.5 POTENTIAL OUTCOME FRAMEWORK

The concept of 'potential outcome' was first developed by Neyman (as mentioned in Rubin, 2005). Ideally, each unit in an experiment should provide two outputs or outcomes simultaneously, one with a treatment allocated to it ($x_T$) and the other without it denoted as a control ($x_C$) so that their difference will be an estimate of the causal effect of $T$. The example given in the previous chapter (Section 7.2.1), about comparing the durability of shoe soles, is a case to remember in this context. Since the experiment relates to shoe, it is possible to have both the observations, $x_T$ and $x_C$, concurrently for a single person.

This is an unusual situation where the potential outcomes for each unit could be available. For most experiments, this will not be so. Only one of the two potential outcomes would be available at a given point of time. For example, for obtaining the effect of a fertilizer on a crop output, it is not possible to have the two potential outcomes simultaneously on the same plot of land. It will have to be either with the fertilizer ($T$) or without it (C). The other will be missing. The important question is how to get the missing value of the potential outcome and hence the causal effect?

Fisher (1935) showed the utility of randomization. It was suggested to have units from a population and allocate them randomly to $T$ and $C$. The random allocation provides the basis for the comparison, without getting explicitly into the problem of the missing values. This has been termed here as the evaluation design with randomization (with cluster or individual units). When randomization is not feasible, one can take recourse to matching to get the missing observations on the potential outcomes.

As mentioned in the previous chapter, a before-after and treatment-control design, wherever possible, can help in strengthening an evaluation design with matching. In many situations, however, this is also not feasible, for instance, for evaluation of a programme that has already been implemented. This may happen if a programme is already in vogue and the measurements on treatment and control groups before implementation of treatment are not available, one is left with the options of having an evaluation design with matching, such as either a treatment-control or a before-after design. The other alternative that can be used is to have an observational study. The procedure of propensity score matching can be fruitfully used in such cases to obtain the potential outcome.

### 8.5.1 Propensity Score Matching

To get a causal relationship, the basic tenet, however, remains that of obtaining the potential outcomes, in other words, to have a comparison between the two groups, $T$ and $C$, by attaining a balance between them in terms of the disturbing variables. The propensity score matching, which helps in attaining the balance has been widely used for this purpose. The method was first proposed by Rosenbaum and Rubin (1983, 1985). This model is popularly known as Rubin Causal Model (Angrist et al., 1996; Holland, 1986). We will discuss briefly to illustrate how allocation can be done to achieve the balance.

The propensity score matching can be used either when an evaluation design with matching is being planned or in a case where data are already available from an observational study. In either case, an idea about the sample size would be required to test the statistical significance of the estimated difference between $T$ and $C$. Using equation 8.1 and following the logic employed in Section 2.2.7, one can arrive at a sample size that would be required to get an estimate of the effect (e.g., the difference in sample means between $T$ and $C$) for a given precision (Lemeshow et al., 1990).

Since matching involves wastage of cases, while deciding the sample size, it is advisable to have extra cases. In an evaluation design with matching, generally $T$ is fixed in advance. The total sample size is then decided taking into account the wastage in matching.

In an observational study, already an available data from a survey could be used for the impact estimation. Since the total available cases that could be used for the matching are already given, no adjustment for the wastage is possible in this case. Matching might reduce the available sample size for the impact estimation. Suppose, the objective is to understand the effect of a vaccine in reducing the incidence of a disease. One can use a large-scale survey on health that offers data on the different socio-economic and other characteristics that could be considered as the disturbing variables for each person along with the following:

**(i)** whether one has suffered from the disease and
**(ii)** whether one has used the vaccination.

One can avoid the matching and use an appropriate statistical model, such as, a regression analysis to obtain the effect of the vaccine. However, these models are generally based on a few assumptions that are often difficult to satisfy. The causal model developed by Rubin can be fruitfully employed in such cases. It is possible, as elaborated subsequently, to use the propensity score matching approach.

Once observations on $T$ and $C$ are obtained, matching or balance between them is obtained by first calculating the propensity score for each observation (Guo and Fraser, 2010). This is achieved by employing a logistic regression, such as

$$y = a + b_1x_1 + b_2x_2 + \cdots + b_kx_k \tag{8.4}$$

where

$$y = \log\left(\frac{T}{1-T}\right)$$

and $T$ is a dichotomous variable taking a value of 1 for a treatment and 0 otherwise (control). $x_i$'s are the disturbing variables and a and $b_i$ are the regression coefficients which can be obtained through fitting equation 8.4. The regression is an attempt to provide an equation based on the disturbing variables that best discriminate between $T$ and $C$. Although there are other techniques that can achieve the same purpose, logistic regression is generally preferred (Guo and Fraser, 2010).

The equation can be used to obtain an overall propensity score for all units belonging to $T$ or $C$, based on the values of the disturbing variables they have. The concept is that if two observations have same propensity score, it can be assumed that they are similarly distributed in terms of the disturbing variables.

Suppose, for evaluation of a public programme in operation in a population, random samples of $T$ and $C$ along with information on their characteristics considered as the disturbing variables are available. Using the data, it is possible to obtain the regression equation 8.4. This will provide the propensity scores for each observation on $T$ and $C$. If two units, one belonging to $T$ and the other to $C$ have similar propensity score, they are balanced in terms of the disturbing variables. From such a pair matching, one can compute the means of $T$ and $C$ to obtain the treatment effect. An example of the procedure is presented below to illustrate the method.

### Example 8.1

A study was undertaken to gauge the effect of watching violent ($T$) compared to non-violent ($C$) computer games on hostility ratings of children ($y$). Following table provides the necessary information, apart from $y$, on:

1. Treatment, $T = 1$ for violent computer games,
   Control, $T = 0$ for non-violent games.
2. Gender of children, $x_1 = 1$ for boys,
   $= 0$ for girls.

3. Work status of mother $x_2 = 1$ if mother is working outside home,
$= 0$ if mother is a housewife.

| Type of computer games ($T$) | Gender ($x_1$) | Mother's work status ($x_2$) | Number of children | Hostility ratings ($y$) | Mean hostility rating ($\bar{y}$) |
|---|---|---|---|---|---|
| 0 | 0 | 0 | 6 | 48,50,52,48,50,52 | 50 |
| 0 | 0 | 1 | 2 | 55,59 | 57 |
| 0 | 1 | 0 | 6 | 53,55,57,53,55,57 | 55 |
| 0 | 1 | 1 | 2 | 60,64 | 62 |
| 1 | 0 | 0 | 2 | 63,67 | 65 |
| 1 | 0 | 1 | 6 | 70,72,74,70,72,74 | 72 |
| 1 | 1 | 0 | 2 | 68,72 | 70 |
| 1 | 1 | 1 | 6 | 75,77,79,75,77,79 | 77 |

For example, there were six girl children ($x_1 = 0$) whose mother was a housewife ($x_2 = 0$) and who watched non-violent games ($T = 0$). Their hostility ratings ($y$) are shown in column 5 and the mean rating is 50. Similarly, there were two girl children ($x_1 = 0$) whose mothers were working outside home ($x_2 = 1$) and who watched non-violent games ($T = 0$) with the mean hostility rating as 57, and so on. The objective is to estimate the effect of watching violent computer games compared to non-violent computer games on children's hostility rating. Gender and mother's work status are the two disturbing variables.

Using the data, one can obtain mean ratings for $T$ and $C$:

Mean hostility rating among

$$T = \frac{[(63+67)+(70+72+74+70+72+74)+(68+72)+(75+77+79+75+77+79)]}{16}$$

$$= 72.75$$

Mean hostility rating among

$$C = \frac{[(48+50+52+48+50+52)+(55+59)+(53+55+57+53+55+57)+(60+64)]}{16}$$

$$= 54.25$$

Therefore, if no adjustment is made for the differential distribution of the two disturbing variables among $T$ and $C$, the effect would be 18.5 (72.75 - 54.25).

It can be checked from the table that even though the distribution of male and female children are same in $T$ and $C$, it is not so for a mother's work status. For example, a much higher proportion of mothers (12 out of 16) of children are working outside among $T$, whereas this is only 4 out of 16 in $C$. Since a mother's work status makes a difference in the hostility ratings (it can be observed from the above table that the mean hostility rating is much higher among children whose mothers are working outside compared to those whose mothers are housewives), this anomaly in the distribution can result in an overestimation of the effect.

To balance the distribution of the disturbing variables between $T$ and $C$, first a logistic regression is run with $P$ (probability that an individual will belong to a treatment) as the dependent variable and $x_1$ and $x_2$ as independent variables, considering all 32 observations. The equation obtained is as follows:

$$\log\left(\frac{P}{1-P}\right) = -1.099 + 0.x_1 + 2.197\,x_2$$

This can be written as follows:

$$P = \frac{\text{Exp}\,(-1.099 + 2.197\;x_2)}{1 + \text{Exp}\,(-1.099 + 2.197\;x_3)} \tag{8.5}$$

With the help of equation 8.5, the value of $P$ which is the propensity score can be computed for each unit. They are estimated and shown in Table 8.1.

**Table 8.1** Estimated values of propensity scores for each child in $T$ and $C$ according to categories of gender and mother's work status.

| $x_1$ | $x_2$ | Propensity score | Individuals in $T$ | Individuals in $C$ |
|---|---|---|---|---|
| 0 | 0 | 0.25 | 63, 67 | 48,50,52,48,50,52 |
| 0 | 1 | 0.75 | 70,72,74,70,72,74 | 55,59 |
| 1 | 0 | 0.25 | 68,72 | 53,55,57,53,55,57 |
| 1 | 1 | 0.75 | 75,77,79,75,77,79 | 60,64 |

It can be checked that, for 16 out of 32 children whose mothers are working outside have a propensity score of 0.75. Out of these 16, 12 children watch violent computer games, that is, they belong to $T$. The remaining 16 children whose mothers are housewives possess a propensity score of 0.25. Of them, 12 belong to $C$ and only 4 to $T$.

In the first category, both $x_1$ and $x_2$ as 0, we have only two children belonging to $T$, whereas there are six in $C$ available for matching. All these six children have their propensity score (same as the two in $T$) and therefore any two can be chosen to match with $T$. In other words, four cases of $C$ will be wasted. The wastages in the other three categories will similarly be four cases each of $T$ from the second, $C$ from the third and $T$ from the fourth. In other words, for this example, there will be 50 per cent wastage from both $T$ and $C$. For the purpose of illustration, we assume that first two units in a category, if having more than two units, will be chosen (the underlined cases in Table 8.1). The eight units chosen in this way for $T$ and $C$ are:

$T$ 63, 67, 70, 72, 68, 72, 75 and 77
$C$ 48, 50, 55, 59, 53, 55, 60 and 64

The mean hostility rate for children in $T$ is 70.5 and for $C$, it is 55.5. The treatment effect, therefore, will be 15 (= 70.5 – 55.5).

The above example helps to illustrate the procedure of obtaining a balance between $T$ and $C$ using the propensity score matching. While planning a study, steps can be taken to minimize the wastage.

Let us consider a situation where an evaluation of a programme that has been in operation for sometime is to be undertaken. The sample size to be considered is, say, $2n$. Generally, $n$ cases of the treatment can be selected randomly from the area where the programme is in operation. Using idea about the wastage, sample size for the control cases will be $n + nk$, where $k$ denotes the percentage of the control cases that would be wasted. If selection of control cases can also be done from the same area where the programme is in operation, the cost of data collection will reduce considerably.

The cluster sampling can be done to select individuals. For example, if the programme refers to an incentive programme to boost education among girl child, it may be worthwhile to consider individual family who accepted the incentive scheme and those who did not accept both belonging to a same PSU (e.g., village). But if the programme to be evaluated refers to the effectiveness of a training programme, selection of control cases in all likelihood should be avoided from the programme area. This is because of the possibility of high spillover effect. After selection of the cases, propensity score matching can be done for obtaining the matched control cases for the $n$ treatment.

There are different procedures of matching. A caliper matching, where matching between $T$ and $C$ is done by examining whether their propensity scores lies within a given interval, is generally used. For example, one could specify, instead of 0.75 (in the above example), an interval between 0.72 and 0.78 for the matching.

## 8.6 CHOICE OF A DESIGN

Having discussed the three basic designs (Chapter 7) and elaborated the methods of allocation, it is pertinent to understand how best to form a design in a specific situation. The strength of a design in terms of its validity and reliability depends on the way the comparison between treatment and control is done in a design and also on the type of allocation of units.

The various evaluation designs can be grouped into the four categories depending mainly on three criteria:

**(i)** timing of the two observations on $T$ and $C$,
**(ii)** possibility of using same (similar) sample for observing both $T$ and $C$, and
**(iii)** ability to assign units randomly to $T$ and $C$.

The four different evaluation designs, discussed in chapter 6, can be formed using the above criteria. These alternative designs along with their advantages and disadvantages are summarized below.

**(a)** Same (similar) samples observed without any time lag
In this case, a treatment-control design will be suitable. Both the selection as well as time effects are controlled and hence the design is internally valid. For example, in the shoe sole experiment (section 7.2) a same sample of individuals provides both

the observations on $T$ and $C$. If same sample is used to obtain the information on treatments, for internal validity, the question on whether the sample is drawn randomly or not is not important. The disturbing variables are perfectly matched between $T$ and $C$. Similarly, in another experiment to compare the effectiveness of a new medicine, a group of patients was randomly allocated to $T$ (with the new medicine) and $C$ (with the existing medicine), discussed in section 10.6.4. In both these cases, the internal validity is ensured. However, the external validity of the design would depend on how the sample of persons in the first case and the group of patients in the second, were selected. Randomization, the most widely recommended method to provide complete protection against the biases, if adopted to select the samples, will additionally ensure generalization to the population from which the selection has been done. If randomization is not used for selecting the sample, caution should be exercised to claim external validity. Such designs, as mentioned in chapter 6, will be considered as evaluation designs with random allocation of units or with clusters assigned randomly to treatment arms.

**(b)** Same (similar) units observed with time lag
If the same units could be observed with a time lag, a before-after design becomes appropriate for evaluating impact. In fact, for evaluation of many intervention strategies (such as door-to-door campaign about institutional delivery (see section 7.2) or marketing strategy to promote a shampoo (see section 7.3.1), a time gap is essential before measuring the output for $T$. There is no selection effect present in this type of design, although one needs to worry about and control the biasing effect of time. As mentioned, if two random samples from a single population are drawn for observing both $T$ and $C$, it will help control the pre-test effect. If other time effects (e.g., history effect, measurement effect etc.) are satisfactorily controlled, the internal validity can be ensured for this evaluation design with random allocation of units (clusters).

**(c)** Different samples with no time lag
A treatment-control design adopted in such cases will have no time effect present. Its effectiveness will depend on the extent to which matching takes care of the disturbing variables. Generally, individuals (clusters) who agree to participate in the treatment group ($T$) are obtained and a corresponding matched group is selected as the control C. It will be an evaluation design with unit or cluster level matching. The moot point is to exercise the control of the disturbing variables. If that can be ensured, it will be as valid as any of the above mentioned designs. In order to improve the effectiveness of a design, particularly if disturbing variables are many in number and difficult to control, the Rubin Causal Model can be fruitfully utilized.

**(d)** Different samples with time lag
A before-after design is possible if there is the possibility of controlling both the selection and the time effects. A before-after and treatment-control design, if feasible, will be ideal in such cases. Otherwise the Rubin Causal Model can be employed to improve the strength of such a design.

In cases where random allocation is ruled out, matching has to be employed to obtain the balance between the two treatment arms. A treatment-control design will be valid only when matching of all the disturbing variables can be ensured. Similarly, a before-after design will be valid only if non-existence of the time effects is possible.

A relatively better design that is recommended is the combination of the two, that is, the before-after and treatment-control design. Although the designs in such cases are regarded as quasi-experimental designs, they offer protection against the biases and are also known as the 'gold standard' among non-experimental designs.

## 8.7 KEY POINTS

- Allocation of units to treatment and control groups is crucial in understanding the treatment effect. In fact, the type of allocation adopted in a design decides its validity as well as reliability.
- Random allocation of units to the two groups is by far the most dependable technique that provides validity as well as reliability to a design. It guards against all the biases.
- The other approach for the allocation, when random assignment is not feasible, is matching. Like randomization, it also tends to provide validity to an experiment. However, unlike the other, which provides protection against all known or unknown effects of the disturbing variables, matching can control only the disturbing variables that are known and explicitly taken into account during allocation of units.
- The two techniques randomization and matching are not mutually exclusive. They are often used together to strengthen a design.
- Matching can be implemented pair-wise. Two units matched in terms of the known disturbing variables can be allocated to the two groups preferably randomly. This is known as pair matching.
- An alternative form of matching is known as stratification. In this case, matching is done in groups rather than in pair.
- When randomization is feasible, a treatment-control design can be both effective and efficient. If the disturbing variables are large, the required sample size will increase. Matching of major disturbing variables can reduce the sample size and make the design more efficient.
- If randomization is not feasible, a before-after and treatment-control design is recommended. In cases where the baseline information on the treatment and control groups are not feasible, Rubin Causal Model can be employed.

# Chapter 9

# Statistical Tests for Measuring Impact

## 9.1 INTRODUCTION

The impact estimation is essentially a comparison of estimates of a characteristic between the two groups denoted as treatment (*T*) and control (*C*). The comparison needs to be tested statistically (by giving allowances for the variations that might have been caused by sampling) for the presence of an impact. For a variable measured on a continuous scale, the parameter to be estimated is usually its mean. For a dichotomous event, the generally preferred parameter to be estimated is a proportion or an odds. Suppose,

$f_T$ = number possessing an attribute among *T*
$f'_T$ = number not possessing the attribute among *T*
$n_1$ = total number of individuals observed in *T*
$= f_T + f'_T$

Similarly,

$f_C$ = number possessing the attribute among *C*
$f'_C$ = number not possessing the attribute among *C*
$n_2$ = Total number of individuals observed in *C*
$= f_C + f'_C$

The observations in treatment and control groups can be represented by a proportion possessing the attribute among *T* ($p_T$) and among *C* ($p_C$), where

$$p_T = \frac{f_T}{n_1}$$

and

$$p_C = \frac{f_C}{n_2}$$

It can also be represented as odds of possessing the attribute among *T* ($O_T$) and *C* ($O_C$), where

$$O_T = \frac{f_T}{f'_T}$$

and

$$O_C = \frac{f_C}{f'_C}$$

$p_T$ signifies what per cent of the individuals observed in $T$ possesses the attribute, whereas $O_T$ reveals the extent to which individuals possessing the attribute is higher compared to its complementary (those not possessing the attribute). Both the indices are related to each other. Although $p_T$ is commonly used, $O_T$ has certain statistical properties that make them preferable.

The risk ratio (RR) and odds ratio (OR) can be written as follows:

$$\text{RR} = \frac{p_T}{p_C} \tag{9.1}$$

$$\text{OR} = \frac{O_T}{O_C} \tag{9.2}$$

### 9.1.1 Two different ways to estimate impact

Let $y_T$ and $y_C$ denote the estimates of a characteristic $y$ in the two groups. There are two measures to assess the impact:

**(a)** Difference $(d) = y_T - y_C$ and
**(b)** Ratio $(r) = \frac{y_T}{y_C}$.

Both these estimates facilitate understanding how $y_T$ compares with that of $y_C$. Suppose, the characteristic $y$ denotes hours of sleep. It is convenient to obtain mean duration of sleep for the comparison, that is, $y_T$ and $y_C$ will signify average hours of sleep for the two groups. For the impact estimation, one can examine their difference ($d$) or a ratio ($r$). When $y$ is a dichotomous variable, $y_T$ and $y_C$ generally refer to either proportion ($p_T$ and $p_C$) or odds ($O_T$ and $O_C$). Either of the two estimates can be used to measure the impact.

It is pertinent to understand the rationale behind choosing between the two measures for the impact estimation. An examination of their characteristics helps in this regard. Both the measures assume that the series of $y$ values, of which only two values are considered for the impact estimation, follows a particular pattern.

The first option connotes a series of $y$ that increases by a fixed magnitude $d$. Any value in the series is expected to be independent of its previous value, that is,

$$y_T = y_C + d$$

In other words, irrespective of the value of $y_C$, $y_T$ is expected to be higher than it by an amount $d$. The increment in a value, in the second case, is expected to depend on the magnitude of the previous value. In other words,

$$y_T = r \times y_C$$

The value of $y_T$ relates to $y_C$ and is, in fact, $r$ times its value. Let us discuss this in the context of the example mentioned in Section 8.3. Suppose per cent of children surviving without antitoxin when brought in moderate condition is 40 ($y_{C1}$) and for the same group with antitoxin is 72 ($y_{T1}$), resulting in a value of $d$ as 32. For the other group, who were brought under severe condition, if per cent surviving without antitoxin is 20, then according to option A, it is expected that among this group, with antitoxin, the survivorship would also increase by 32, that is, it should be 52. Suppose, it is not 52, but 36.

The interpretation, under the assumption of a linear increase, would be that (because the two $d$ values, 32 and 16 are not the same) there is an interaction. That is, if the two $d$ values are statistically different, it will be deemed to be the presence of an interaction or that the effect of antitoxin changes depending on the condition of a patient at the time of administration of the antitoxin.

What if we change the assumption from a linear to a geometric growth as in the second case? The value of $OR$ among the first group (with a moderate condition) is 1.8 $\left(=\frac{72}{40}\right)$ and it is also the same for the second group $\left(=\frac{36}{20}\right)$. Hence, according to this index, the interpretation would be that the effect of the antitoxin is to the extent of increasing the survival chance by 80 per cent among patients given antitoxin compared to those not given antitoxin. This effect is same, irrespective of whether a patient is administered the treatment in moderate or severe condition.

Therefore, as it is evident, the interpretation of an effect, particularly the interaction effect, can depend on a particular index being employed for the impact estimation (Breslow and Day, 1980).

Since there are only two observations, it is not possible to provide further idea about whether the assumption of a linear or a geometric growth would better describe the relationship (increase in the output). For this, at least three categories are needed. For the impact analysis, the focus here is on measuring the main effect (equation 7.1). Whenever doubts about a specific model to use arise, it would be better to check the impact considering both the indices.

The statistical tests for assessment of an impact are discussed separately considering units of allocation (unit or clusters of units) and for the three types of allocation:

**(i)** completely random,
**(ii)** random within each stratum and
**(iii)** pair matching.

## 9.2 IMPACT WHEN UNITS ARE ALLOCATED RANDOMLY

Generally, sampling variance of an estimate is required to obtain the statistical significance of a treatment effect. Discussions are presented separately for testing an impact when means, risk ratios and odds ratios are used as the estimates. The sampling variance of means is already presented in Chapter 7. For proportions and odds it will be presented subsequently.

### 9.2.1 Testing Difference between Two Means

For testing difference between two means, one for the treatment ($\bar{y}_T$) and the other for control ($\bar{y}_C$), a $t$-statistic can be used, if the sample sizes are small (less than 30). If the two sample sizes are denoted as $n_1$ and $n_2$ with $S_C^2$ and $S_T^2$ being the element variance of the two samples, then, under the null hypothesis that

$$H_0 : \bar{y}_T = \bar{y}_C$$

and against the alternative hypothesis that $\bar{y}_T$ is not equal to $\bar{y}_C$,

$$t = \frac{\bar{y}_T - \bar{y}_C}{S\sqrt{\frac{1}{n_1} + \frac{1}{n_2}}} \tag{9.3}$$

follows a $t$-distribution, where

$$S^2 = \frac{(n_1 - 1)s_C^2 + (n_2 - 1)s_T^2}{n_1 + n_2 - 2} \tag{9.4}$$

is the combined element variance of the two groups.

The numerator of $t$ is the difference between the two means and the denominator gives its standard error. Larger the element variances in $T$ and $C$, greater will be the error; and, larger the two sample sizes or the degrees of freedom (denoted as $df$), which is equal to $n_1 + n_2 - 2$, less will be the error. If null hypothesis is true, then $t$ is expected to be close to zero. Some deviations from the null value are expected because of the sampling fluctuations. The standard error of the difference between the two estimates provides an idea about the extent of such variations. Consequently, for a given value of the denominator, the chance of obtaining a large value of $t$ should be small. For different values of $df$, a table is available (Fisher and Yates, 1963) that shows the probability of obtaining a particular value of $t_{\alpha/2}$ or higher, where $1 - \alpha$ is the probability or the level of confidence that there is no evidence against the null hypothesis.

To illustrate, with a value of $df$ as 20 in an experiment, the values of $t_{0.025}$ and $t_{0.005}$ are 2.086 and 2.845, respectively (Fisher and Yates, 1963). This means, under the null hypothesis there is only a 5 per cent chance of getting a value of $t$ equal to 2.086 or higher. In other words, there is a 95 per cent chance, if the value of $t$ is less than 2.086, that there is no evidence against the null hypothesis of equality between the two means and hence it should not be rejected. If one wants to be 99 per cent confident, one should check similarly with $t_{0.005}$, which is 2.845. It can be checked from the tabulated value of the $t$-statistic that as sample size increases and the standard error decreases, the range of permissible departure from the null value reduces, albeit slowly. When the degree of freedom is 60, the critical value of $t_{0.025}$ is 2.000 and that of $t_{0.005}$ is 2.660.

In order to apply the $t$-distribution, there are a few conditions that should be satisfied. They are:

1. the two populations from which samples $y_T$ and $y_C$ are selected should follow approximately a normal distribution,

2. the two samples should be selected independently and
3. the two population variances are equal.

If allocation to $T$ and $C$ are made randomly from a population and particularly if sample sizes are equal, the assumptions pose no problem. It needs to be mentioned that in an evaluation design with matching, it is generally assumed that all the biasing effects are eliminated and the difference between the two estimates, under the null hypothesis, will be an error that would be randomly distributed. However, for the estimated sampling variance of $T$ and $C$, particularly if the two sample sizes are not similar, the assumption of equality of variances may not hold. The statistical test mentioned above will then require modification.

#### 9.2.1.1 Testing means from two different populations

If the sample sizes are not similar and the condition of equality of variances also does not hold, one can use

$$t = \frac{\bar{y}_T - \bar{y}_C}{\sqrt{\frac{s_C^2}{n_1} + \frac{s_T^2}{n_2}}} \tag{9.5}$$

as an approximation to the $t$-statistic. There is a need to adjust the $df$ in this case (Satterthwaite, 1946). The degrees of freedom (denoted as $v$), instead of $n_1 + n_2 - 2$, is estimated as

$$v = \frac{\left(\frac{s_C^2}{n_1} + \frac{s_T^2}{n_2}\right)^2}{v_1 + v_2} \tag{9.6}$$

where

$$v_1 = \frac{\left(\frac{s_C^2}{n_1}\right)^2}{(n_1 - 1)} \quad \text{and} \quad v_2 = \frac{\left(\frac{s_T^2}{n_2}\right)^2}{(n_2 - 1)}$$

The fractional part of $v$, if any, is disregarded and the integer value is taken as the degree of freedom to check, as indicated earlier, with the tabulated value of the $t$-statistic.

If the design used is a 'before-after and treatment-control' design, there will be four groups of observations: one on control and treatment before application of the treatment and the same two groups after application of the treatment. Let the sample sizes for them be $n_1$, $n_2$, $n_3$ and $n_4$, respectively. Also, let $s_{C1}^2$, $s_{T1}^2$, $s_{C2}^2$ and $s_{T2}^2$ be their respective element variances. Then the value of the $t$-statistic will be

$$t = \frac{(\bar{y}_{T2} - \bar{y}_{T1}) - (\bar{y}_{C2} - \bar{y}_{C1})}{\sqrt{\frac{s_{C1}^2}{n_1} + \frac{s_{T1}^2}{n_2} + \frac{s_{C2}^2}{n_3} + \frac{s_{T2}^2}{n_4}}} \tag{9.7}$$

#### 9.2.1.2 Large-sample *z*-test

If sample sizes are large, both $n_1$ and $n_2$ being greater than 30, the *t*-distribution converges to a normal distribution, denoted by *z*, a standard normal variate, having a mean equal to zero and standard error of unity. It can be expressed as

$$Z = \frac{\bar{y}_T - \bar{y}_C}{\sqrt{\frac{s_C^2}{n_1} + \frac{s_T^2}{n_2}}} \tag{9.8}$$

If a value of $z_{0.025}$ is greater than 1.96, it signifies that the two means are statistically different from each other and the null hypothesis of equality between the two is rejected at the 95 per cent confidence level. It may be noted that for a *t*-test, the value of $t_{0.025}$ is 2.00 with 60 degrees of freedom, which is quite close to $z_{0.025}$. Since the sample sizes are large, both $s_T^2$ and $s_C^2$ are good estimates of the corresponding population variances. The condition of equality of variances can then be relaxed.

### Example 9.1

The objective of a study is to understand whether the household size varies among people who possess agricultural land (*T*) compared to those who do not possess any such land (*C*). Utilizing the data shown in Example 2.1, examine whether the difference between *T* and *C* is statistically significant under the following two assumptions:

1. the two samples *T* and *C* are drawn randomly from the same population and
2. the two samples are drawn from two different populations.

For convenience, the data on *T* and *C* are presented separately below:

| Treatment | Household size | | | | | | | | | | | | |
|---|---|---|---|---|---|---|---|---|---|---|---|---|---|
| *T* | 5 | 6 | 4 | 6 | 5 | 2 | 6 | 6 | 4 | 7 | 6 | 7 | 8 |
| | 5 | 12 | 5 | 7 | 3 | 4 | 6 | 5 | 5 | 3 | 6 | 4 | |
| *C* | 4 | 2 | 3 | 5 | 4 | 2 | 4 | 6 | 2 | 5 | 4 | 1 | 5 |
| | 6 | 8 | 8 | 7 | 7 | 4 | 6 | 3 | 6 | 5 | 3 | 5 | |

$$n = 25,\ \bar{y}_T = \frac{137}{25} = 5.48 \text{ and } \bar{y}_C = \frac{115}{25} = 4.60,$$

$$s_T^2 = \frac{(5^2 + 6^2 + \cdots + 4^2) - 25 \times 5.48^2}{24} = 3.84,$$

$$s_C^2 = \frac{(4^2 + 2^2 + \cdots + 5^2) - 25 \times 4.60^2}{24} = 3.58.$$

Under the assumption that both are random samples from the same population,

$$s^2 = \frac{24 \times 3.84 + 24 \times 3.58}{(25 + 25 - 2)} = 3.71 \text{ and } s = \sqrt{3.71} = 1.93.$$

Using equation 9.3,

$$t = \frac{5.48 - 4.60}{1.93 \times 0.28} = 1.61, \text{ with } df = 48.$$

The critical value of $t_{.025}$ corresponding to 48 degrees of freedom is 2.01. Hence, there is no evidence against the null hypothesis that the two means are from the same population. In other words, the mean household sizes between those having agricultural land (5.48) or those who do not possess any agricultural land (4.60) are not statistically different from each other.

Under the assumption that the two samples are from different populations, the overall degrees of freedom can be calculated as follows:

$$v_1 = \frac{\left(\frac{3.58}{25}\right)^2}{24} = 0.00086 \text{ and } v_2 = \frac{\left(\frac{3.84}{25}\right)^2}{24} = 0.00098$$

and using equation 9.6, $v = 47$. From equation 9.5, $t = 1.61$. The interpretation will remain same as above.

### 9.2.1.3 Testing several means: Application of ANOVA

The analysis of variance (ANOVA) technique can be suitable, particularly when there are more than two means to be tested for the equality. Let $y_{ij}$ be the $j$th ($j = 1, 2, \ldots, n_i$) observation belonging to the $i$th ($i = 1, 2, \ldots, m$) treatment. The set of observations can be represented as follows:

| No. of treatments | S. No. of observations | | | | Total | Mean |
|---|---|---|---|---|---|---|
| | 1 | 2 | $j$ | $n_i$ | | |
| 1 | $y_{11}$ | $y_{12}$ | $y_{1j}$ | $y_{1n_1}$ | $y_1$ | $\bar{y}_1 = \frac{y_1}{n_1}$ |
| 2 | $y_{21}$ | $y_{22}$ | $y_{2j}$ | $y_{2n_2}$ | $y_2$ | $\bar{y}_2 = \frac{y_2}{n_2}$ |
| … | | | | | | |
| … | | | | | | |
| $i$ | $y_{i1}$ | $y_{i2}$ | $y_{ij}$ | $y_{in_i}$ | $y_i$ | $\bar{y}_i = \frac{y_i}{n_i}$ |
| $m$ | $y_{m1}$ | $y_{m2}$ | $y_{mj}$ | $y_{mn_m}$ | $y_m$ | $\bar{y}_m = \frac{y_m}{n_m}$ |

Total number of observations = $n = n_1 + n_2 + \cdots + n_m$. As the name of the test suggests, it analyses the variation in the observations, $y_{ij}$. It is pertinent to note that:

**(i)** a difference, if any, between $y_{ij}$ and $y_{ij'}$ for a fixed treatment $i$), is expected to tell us about the error, because they are random samples from the same population and
**(ii)** a difference between $y_{ij}$ and $y_{i'j}$ can arise because of a difference in the treatment as well as error.

In other words, the total variations among $y_{ij}$, which is expressed as the square of difference between each $y_{ij}$ from the overall mean, can be subdivided into two categories: a) that due to a difference between the treatments and b) that due to error.

The total variance or the total sum of square (TSS) can be represented as

$$\text{TSS} = \Sigma_i \Sigma_j (y_{ij} - \bar{y})^2. \tag{9.9}$$

Alternatively, TSS can also be computed as

$$\text{TSS} = \Sigma_{i=1}^{m} \Sigma_{j=1}^{n_i} y_{ij}^2 - \frac{G^2}{n} \tag{9.9a}$$

where

$$\bar{y} = \frac{\Sigma_{i=1}^{m} \Sigma_{j=1}^{n_i} y_{ij}}{n} \text{ and } G = \Sigma_{i=1}^{m} \Sigma_{j=1}^{n_j} y_{ij}$$

The total variation can also be written as

$$\text{TSS} = \Sigma_{i=1}^{m} \Sigma_{j=1}^{n_i} (y_{ij} - \bar{y}_i + \bar{y}_i - \bar{y})^2$$

$$= \Sigma_{i=1}^{m} \Sigma_{j=1}^{n_i} (y_{ij} - \bar{y}_i)^2 + \Sigma_{i=1}^{m} \Sigma_{j=1}^{n_i} (\bar{y}_i - \bar{y})^2 + 2\Sigma_{i=1}^{m} \Sigma_{j=1}^{n_i} (y_{ij} - \bar{y}_i)(\bar{y}_i - \bar{y})$$

The product term

$$= 2\Sigma_{i=1}^{m} \Sigma_{j=1}^{n_i} (y_{ij} - \bar{y}_i)(\bar{y}_i - \bar{y})$$

becomes zero (Snedecor and Cochran, 1989). The TSS can, therefore, be written as

$$\text{TSS} = \text{Tt.SS} + \text{ESS} \tag{9.10}$$

where

Tt.SS = sum of square due to treatments (treatment sum of squares)

$$= \Sigma_{i=1}^{m} \Sigma_{j=1}^{n_i} (\bar{y}_i - \bar{y})^2 = \Sigma_i n_i (\bar{y}_i - \bar{y})^2 \tag{9.11}$$

Alternatively, it can be calculated as

$$\text{Tt.SS} = \Sigma_{i=1}^{m} \frac{y_i^2}{n_i} - \frac{G^2}{n} \tag{9.11a}$$

where

ESS = error sum of square

$$= \sum_{i=1}^{m} \sum_{j=1}^{n_i} (y_{ij} - \bar{y}_i)^2 \tag{9.12}$$

The calculation of TSS (equation 9.9a) and Tt.SS (equation 9.11a) are straightforward, and the value of ESS is obtained from their difference (equation 9.10). A large value of Tt.SS denotes that the treatment means are differing from each other and so also from the overall mean. Larger its value with respect to ESS, greater is the chance that the treatments differ significantly from each other.

On the other hand, a large value of ESS will mean that the population from which the samples are selected is heterogeneous. Chances of getting a significant impact of treatment will reduce. As has been discussed earlier, the error or ESS can be reduced either by increasing the sample size or changing the design. Stratification, which is also known as blocking, can also be used to reduce the heterogeneity (Chapters 2 and 8).

Under the null hypothesis that

$$H_0 : \bar{Y}_1 = \bar{Y}_2 = \cdots = \bar{Y}_m,$$

Where $\bar{Y}_i$ is the mean value for the characteristic $y$ in the case of the $i$th treatment in the population, the ratio of the two sum of squares

$$F = \frac{\dfrac{\text{Tt.SS}}{(m-1)}}{\dfrac{\text{ESS}}{(n-m)}}$$

follows $F$-distribution with $(m - 1)$ and $(n - m)$ degrees of freedom. The factor $(m - 1)$ is the degree of freedom for the Tt.SS (because there are $m$ treatments) and since TSS has $(n - 1)$ degrees of freedom, the ESS has $n - m$ $(= n - 1 - m + 1)$ degrees of freedom. Under the null hypothesis, the value of $F$ is expected to be reasonably small and around a value of 1. The chance of getting a value far greater than 1 is, however, expected to be small. For example, if there are three treatments ($m = 3$) and all together 63 observations, the tabulated value of $F_{2, 60}$, corresponding to a chance of 0.05 (or a 95 per cent confidence), is 3.15. That is, if the null hypothesis is true, the chance of getting a value of $F$ of 3.15 or higher is small and less than 5 per cent. This means that if the calculated value of $F$ is 3.15 or higher, one can be 95 per cent confident in rejecting the null hypothesis of equality between the treatment means. Let the three treatments be denoted as $C$, $T_1$ and $T_2$, that is, apart from the control, there are two different treatments. The rejection of the null hypothesis would only indicate that at least one of the treatments is different from the others. It would not provide any idea about the relative effectiveness of $T_1$ and $T_2$. Such inferences are also possible (McClave et al., 2005). The ANOVA test also depends on few assumptions and they are similar to the three mentioned in the case of the $t$-test.

#### 9.2.1.4 Non-parametric tests

The statistical tests of the difference between two or more means, discussed so far assumes that the population under study follows a parametric distribution, namely, the normal distribution. No such assumption is required in the case of a non-parametric test. In the following, two frequently used tests have been discussed.

**(a) Wilcoxon rank sum test**

The Wilcoxon rank sum test is a suitable test to check whether two samples are from the same population or belong to two different populations. Let $y_{ci}$ $(i = 1, 2, \ldots, n_1)$ and $y_{Ti}$ $(i = 1, 2, \ldots, n_2)$ denote the $n_1$ and $n_2$ values of the characteristics (say hours of sleep to test the effect of a sleeping pill) under study in the control and treatment groups. The two samples $n_1$ and $n_2$ are taken together and the values are ranked from the lowest to the highest (1 to $n_1 + n_2$). While ranking, it is possible to find two or more of the $y$ variables having same values. For such tied cases, their average rank value is substituted (further elaborated in Example 9.2).

The basic premise of the method is simple. If the treatment, sleeping pill, in this case, is effective, $y_{Ti}$ will, in general, be higher than $y_{Ci}$. In other words, ranks of the variables belonging to $T$ will, in general, be higher than that in $C$. However, if the treatment does not have an effect, the ranks will be randomly distributed between the two groups and their rank sums would tend to be similar. Suppose,

$\text{Sum}_T$ = aggregates of the ranks of variables belonging to $T$
$\text{Sum}_C$ = aggregates of the ranks of variables belonging to $C$

It can be checked that

$$\text{Sum}_T + \text{Sum}_C = \frac{n(n+1)}{2}$$

where

$$n = n_1 + n_2$$

If an effect is reflected by increasing values of $y$, a treatment effect is expected to result in $\text{Sum}_T$ to be greater than $\text{Sum}_C$. For example, the impact to be evaluated might refer to, say, per cent surviving among those exposed to a specific medicine (the treatment, $T$) in comparison to the control $C$, who are not exposed to the medicine. Similarly, the opposite will be true if the output ($y$) considered is per cent dying, instead of surviving, that is, if effective, $\text{Sum}_T$ will be smaller than $\text{Sum}_C$.

For small sample sizes, when both $n_1$ and $n_2$ are less than 10, tables are available to take a decision regarding whether the two rank sums are so different as to consider them to belong to two different populations (i.e., treatment is significantly different from the control). For different values of $n_1$ and $n_2$, the table presents a lower and an upper limit for the smaller value of the rank sum, that is, $\text{Sum}_C$ (McClave et al., 2005).

For larger sample sizes, both $n_1$ and $n_2$ greater than 10, it has been shown that, under the null hypothesis of equality of the two distributions

$$E(\text{Sum}_C) = \mu_C = \frac{n_1(n+1)}{2} \tag{9.13}$$

and

$$V(\text{Sum}_C) = \sigma_C^2 = \frac{n_1 n_2 (n+1)}{12} \tag{9.14}$$

In other words, $\text{Sum}_C$ can be assumed to follow a normal distribution with mean, $\mu_C$ and variance $\sigma_C^2$. Consequently,

$$z = \frac{\text{Sum}_C - \frac{n_1(n+1)}{2}}{\sqrt{\frac{n_1 n_2 (n+1)}{12}}} \tag{9.15}$$

is a standard normal variate. If the value of $z$ is greater than 1.96, it will mean that the null hypothesis of equal population distribution of $C$ and $T$ are not tenable and can be rejected. It can be said that the chance of getting such a value of $z$, under the null hypothesis, is ≤0.05.

If $\text{Sum}_T$ happens to be smaller than $\text{Sum}_C$, then

$$z = \frac{\text{Sum}_T - \frac{n_2(n+1)}{2}}{\sqrt{\frac{n_1 n_2 (n+1)}{12}}} \tag{9.16}$$

should be used to test whether the value of the standard normal variate is greater than 1.96.

The two samples are assumed to be independently and randomly drawn. There is no restriction imposed on the nature of the two population distributions. However, the variable under study needs to be measured in a continuous scale. Its use is avoided if a variable is discrete. This is because for a discrete variable, the number of ties, which occurs while ranking when two or more of them have same values, can be large (McClave et al., 2005).

**(b) Kruskal–Wallis *H*-test**

This is another non-parametric test which can be employed to test whether two or more samples drawn independently and randomly can be considered to belong to the same population or not. Let there be $m$ treatments and $y_{ij}$ denote the $j$th ($j = 1, 2, ..., n_j$) observation representing $i$th ($i = 1, 2, ..., m$) treatment. The total sample size is $n$, where

$$n = n_1 + n_2 + \cdots + n_m.$$

The null hypothesis to be tested is

$$H_0 = \bar{Y}_1 = \bar{Y}_2 = \cdots = \bar{Y}_m$$

where $\bar{Y}_i$ is the mean of the $i$th treatment. Like in Wilcoxon method, the $m$ samples are pooled together and ranked according to $y_{ij}$ from the lowest (having a rank of 1) to the highest (with a rank of $n$). The tied cases are as usual given an average rank. Let $r_{ij}$ denote the rank of $j$th observation in the $i$th sample. The total rank sum will be

$$r = \frac{n(n+1)}{2}$$

If $r_i$ is the rank sum total of cases in $i$th treatment, the statistic $H$ is given as

$$H = \frac{12}{n(n+1)} \Sigma_i \frac{r_i^2}{n_i} - 3(n+1) \tag{9.17}$$

Under the null hypothesis, since the $m$ samples are assumed to have been allocated randomly to each treatment, ideally (but for the sampling fluctuations) $r_j$ will be given by

$$r_i = \frac{r}{m} = \frac{n(n+1)}{2m}$$

It can easily be checked that under such ideal conditions (substituting the values of $r_i$ in equation 9.17), $H$ will be equal to zero. It has been shown that $H$, under the null hypothesis, follows a chi-square distribution with $m - 1$ degrees of freedom, and the rejection or otherwise of the null hypothesis can be checked accordingly. Here too, like in the previous test, the output variable is assumed to have been measured in a continuous scale.

### Example 9.2

A study was undertaken to examine the impact of an incentive scheme to raise the age at marriage among females in a population. Following data shows age at marriage of girls who got married during last 1 year according to their level of education (in two categories 0, if educated less than secondary, and 1, if educated beyond secondary) and whether a family joined the incentive scheme ($T$) or did not join it ($C$).

| Age at marriage | | | | | |
|---|---|---|---|---|---|
| Education level less than secondary | | | Education level secondary or above | | |
| S. No. | $T$ | $C$ | S. No. | $T$ | $C$ |
| 1 | 16.75 | 13.92 | 1 | 19.17 | 17.92 |
| 2 | 18.42 | 12.33 | 2 | 18.50 | 18.00 |
| 3 | 18.67 | 12.50 | 3 | 18.75 | 20.67 |
| 4 | 13.67 | 12.25 | 4 | 18.83 | 20.33 |
| 5 | 21.25 | 13.83 | 5 | 23.17 | 20.83 |
| 6 | 16.33 | 13.08 | 6 | 23.58 | 21.75 |
| 7 | 18.25 | 13.25 | 7 | 23.33 | 20.17 |
| 8 | 18.25 | 14.17 | 8 | 25.00 | 14.33 |
| 9 | 17.00 | 14.42 | 9 | 18.33 | 19.25 |
| 10 | 16.42 | 15.17 | 10 | 21.83 | 22.00 |
| 11 | 16.50 | 15.25 | 11 | 23.67 | 17.67 |
| 12 | 21.33 | 16.08 | 12 | 23.25 | 18.08 |
| 13 | 21.92 | 16.17 | 13 | 24.42 | 16.42 |

| 14 | 18.25 | 17.17 | 14 | 19.33 | 20.58 |
|---|---|---|---|---|---|
| 15 | 19.08 | 17.25 | 15 | 19.67 | 19.42 |

**ANOVA test**

Although there are only two categories of the treatment, it is possible to apply an ANOVA test to examine whether there is a significant treatment effect or whether the mean age at marriage among $T(\bar{y}_T)$ is same or higher compared to that of $C(\bar{y}_C)$.

$$G = [(16.75 + 18.42 + \cdots + 19.33 + 19.67) + (13.92 + 12.33 + \cdots + 20.58 + 19.42)]$$

$$= (504.26 + 592.92) = 1097.18, \frac{G^2}{n} = \frac{1097.18^2}{60} = 20063.4$$

$$\text{Tt.SS} = \frac{504.26^2}{30} + \frac{592.92^2}{30} - 20063.4 = 131.01$$

$$\text{TSS} = (13.92^2 + 12.33^2 + \cdots + 20.58^2 + 19.42^2) - 20063.4$$

$$= 20694.0 - 20063.4 = 630.60$$

$$\text{ESS} = \text{TSS} - \text{Tt.SS} = 499.59, \text{df for ESS} = 60 - 2 = 58 \text{ and df for the Tt.SS} = 2 - 1 = 1$$

Therefore,

$$F = \frac{\frac{131.01}{1}}{\frac{499.59}{58}} = 15.21$$

which is highly significant at 1 and 58 degrees of freedoms. The null hypothesis of equality between the two means is thus rejected, implying that the mean age at marriage among *T* is higher compared to *C* even after taking care of the sampling fluctuations.

**Wilcoxon rank sum test**

Ignoring the level of education, age at marriage among all the 60 females belonging to *T* and *C* are considered together and ranked from 1 to 60 as shown below:

| | **Values of age at marriage and their ranks (in parenthesis)** | | | | | | | | | |
|---|---|---|---|---|---|---|---|---|---|---|
| For units in *T* | 1 | 2 | 3 | 4 | 5 | 6 | 7 | 8 | 9 | 10 |
| | 16.75 (20) | 18.42 (32) | 18.67 (34) | 13.67 (6) | 21.25 (48) | 16.33 (16) | 18.25 (29) | 18.25 (29) | 17.00 (21) | 16.42 (17.5) |
| | 11 | 12 | 13 | 14 | 15 | 16 | 17 | 18 | 19 | 20 |
| | 16.50 (19) | 21.33 (49) | 21.92 (52) | 18.25 (29) | 19.08 (37) | 19.17 (38) | 18.50 (33) | 18.75 (35) | 18.83 (36) | 23.17 (54) |
| | 21 | 22 | 23 | 24 | 25 | 26 | 27 | 28 | 29 | 30 |
| | 23.58 (57) | 23.33 (56) | 25.00 (60) | 18.33 (31) | 21.83 (51) | 23.67 (58) | 23.25 (55) | 24.42 (59) | 19.33 (40) | 19.67 (42) |

| For units in C | 1 | 2 | 3 | 4 | 5 | 6 | 7 | 8 | 9 | 10 |
|---|---|---|---|---|---|---|---|---|---|---|
| | 13.92 (8) | 12.33 (2) | 12.50 (3) | 12.25 (1) | 13.83 (7) | 13.08 (4) | 13.25 (5) | 14.17 (9) | 14.42 (11) | 15.17 (12) |
| | 11 | 12 | 13 | 14 | 15 | 16 | 17 | 18 | 19 | 20 |
| | 15.25 (13) | 16.08 (14) | 16.17 (15) | 17.17 (22) | 17.25 (23) | 17.92 (25) | 18.00 (26) | 20.67 (46) | 20.33 (44) | 20.83 (47) |
| | 21 | 22 | 23 | 24 | 25 | 26 | 27 | 28 | 29 | 30 |
| | 21.75 (50) | 20.17 (43) | 14.33 (10) | 19.25 (39) | 22.00 (53) | 17.67 (24) | 18.08 (27) | 16.42 (17.5) | 20.58 (45) | 19.42 (41) |

It can be seen that age at marriage of 16.42 years occur twice. If they were not exactly equal but occurred with a slight difference, their ranks would have been 17 and 18. Since there is a tie, they both are assigned a rank of 17.5 (average of 17 and 18). Similarly, the value 18.25 occurred thrice. They are assigned (29) each, which is average of 28, 29 and 30.

$$\text{Sum}_C = 8 + 2 + \cdots + 45 + 41 = 686.5$$
$$\text{Sum}_T = 20 + 32 + \cdots + 40 + 42 = 1143.5$$
$$n\,\frac{(n+1)}{2} = 1830.$$

Since the effect is shown by age at marriage ($y$) which increases in the presence of an effect, equations 9.13–9.15 are used to get the effect.

$$E(\text{Sum}_C) = \frac{30\times 61}{2} = 915 \text{ and } V(\text{Sum}_C) = \sigma_C^2 = \frac{30\times 31\times 61}{12} = 4575. \text{ Hence, } z = -3.38$$

The chance of obtaining such a low value of $z$ is very small and highly significant under the null hypothesis that the two samples of $T$ and $C$ are from the same population. The null hypothesis is, therefore, rejected. This indicates that age at marriage is in general, higher among $T$ compared to $C$ or there is a significant increase in age at marriage among those who accepted the incentive.

**Kruskal-Wallis test**

$$r = \frac{n(n+1)}{n} = \frac{60\times 61}{2} = 1830$$
$$r_1 = \text{rank Sum in } T = \text{Sum}_T = 1143.5,\ r_2 = \text{Sum}_C = 686.5$$
$$H = \frac{12}{60\times 61}\left(\frac{1143.5^2}{30} + \frac{686.5^2}{30}\right) - 3\times 61 = 194.41 - 183 = 11.41.$$

According to the tabulated value of chi-square with 1 degree of freedom at 5 per cent level is 3.84. The chance of obtaining a value of 11.41, if null hypothesis of equality between the two means is true, is much smaller and untenable.

### 9.2.2 Testing Difference Between Two Proportions

When a characteristic under study is dichotomous, a proportion is used to understand the extent of prevalence of an event. Two proportions, one for the treatment group ($p_T$) and the other for the control group ($p_C$) can be compared using $t$ or normal distribution, as mentioned for the means. The variance of the difference is given by

$$V(p_T - p_C) = \frac{p_T q_T}{n_2} + \frac{p_C q_C}{n_1}$$

where

$$q_T = 1 - p_T \text{ and } q_C = 1 - p_C.$$

#### 9.2.2.1 Chi-square test of independence

Another popular test for understanding the significance of difference between two proportions is the chi-square test. When units are allocated to treatments randomly, a chi-square test of independence between the two groups is suitable. Let us consider the estimation of impact of a vaccination programme in preventing the occurrence of disease in a population. A group of $n$ individuals selected randomly from the population can be subdivided as follows:

| Event | Control | Treatment | Total |
|---|---|---|---|
| Disease occurred | $f_C$ | $f_T$ | $f$ |
| Did not occur | $f_C'$ | $f_T'$ | $f'$ |
| Total | $n_1$ | $n_2$ | $n$ |

The test function is formed by obtaining the expected number of cases in the four cells. Let $E(f_C)$ be the expected number of persons, out of $n$, who have not received the vaccination (in C) and have suffered from the disease. Therefore,

$$E(f_C) = n \cdot P(f_C)$$

where P ($f_C$) is the probability that a randomly selected person will be found to have not received the vaccine and has suffered from the disease. $P(f_C)$ is the product of two different probabilities. One is the probability that a selected person will be found to have not received the vaccine ($P(C)$) and the probability that a person not vaccinated will suffer from the disease ($P(f)$). Now, since units are allocated randomly to treatment and control,

$$P(C) = \frac{n_1}{n}$$

Under the null hypothesis of equal chance of suffering from the disease among both vaccinated and non-vaccinated groups, we have

$$\frac{f_C}{n_1} = \frac{f_T}{n_2} = \frac{f}{n}$$

Therefore,

$$P(f) = \frac{f_C}{n_1} = \frac{f}{n}$$

In other words,

$$P(f_C) = \frac{n_1}{n} \cdot \frac{f}{n}$$

and

$$E(f_C) = n_1 \cdot \frac{f}{n}$$

In a similar manner,

$$E(f_T) = n_2 \cdot \frac{f}{n}$$

$$E(f_{C'}) = n_1 \cdot \frac{f'}{n}$$

$$E(f_{T'}) = n_2 \cdot \frac{f'}{n}.$$

It has been shown that under the null hypothesis,

$$\chi^2 = \frac{(f_C - E(f_C))^2}{E(f_C)} + \frac{(f_{C'} - E(f_{C'}))^2}{E(C')} + \frac{(f_T - E(f_T))^2}{E(f_T)} + \frac{(f_{T'} - E(f_{T'}))^2}{E(f_{T'})} \quad (9.18)$$

follows a chi-square distribution with 1 degree of freedom. The degrees of freedom is calculated as

$$\text{Degrees of freedom} = (l - 1)(m - 1)$$

where $m$ is the number of treatment and $l$ denotes the number of categories of the event. Since both of them are equal to 2, in the present example (referred to as a $2 \times 2$ table), the degree of freedom is 1.

If null hypothesis is true, the value of $\chi^2$, if the sampling fluctuations can be ignored, will be equal to zero. The tabulated values of $\chi^2$ at different degrees of freedom are available in most of the textbooks on statistics (Fisher and Yates, 1963). They help in deciding whether the value of $\chi^2$ is large enough to reject the null hypothesis of equality between the two proportions. For example, if null hypothesis is true, the chance of getting a value of $\chi^2$ at 1 degree of freedom which is equal to or greater than 3.84 is $\leq$5 per cent. Hence, if the value of $\chi^2$ computed from equation 9.18 is greater than or equal to 3.84, there is a reason to infer that the null hypothesis be rejected, which would imply that the two treatments are different from each other.

Equation 9.18 can be easily extended if there are more than two treatments and more than two categories for an output variable (Yule and Kendall, 1950). Equation 9.18, however, requires little modification when the test is carried out for a 2 × 2 table. This is particularly true if the number of observations for the analysis is small (it is suggested that the expected value of each cell should be at least 5). The modification requires that the numerator in equation 9.18 should be changed as

$$\chi^2 = \frac{(|f_C - E(f_C)| - 0.5)^2}{E(f_C)} + \frac{(|f_{C'} - E(f_{C'})| - 0.5)^2}{E(f_{C'})} + \frac{(|f_T - E(f_T)| - 0.5)^2}{E(f_T)} + \frac{(|f_{T'} - E(f_{T'})| - 0.5)^2}{E(f_{T'})} \quad (9.19)$$

That is, an additional deduction of 0.5 is carried out from each value of the absolute difference between an observation in a category and its expected value. The convergence to a chi-square distribution is only approximate in a 2 × 2 table and hence it is advisable to make the corrections (Yates, 1934).

#### 9.2.2.2 Testing odds ratio

The odds ratio can also serve the purpose of examining whether there is a difference between $T$ and $C$. In fact, the use of odds ratio also helps in gauging the strength of a relationship, if $T$ differs from $C$. The odds ratio, that is, the ratio of odds in treatment to that in the control group is

$$\text{Odds ratio of the disease} = \frac{\dfrac{f_T}{f_{T'}}}{\dfrac{f_C}{f_{C'}}} = \frac{f_T f_{C'}}{f_{T'} f_C} \quad (9.20)$$

It shows the extent to which the odds of disease (ratio of those with and without the disease) differ in $T$ compared to that in $C$. If the odds ratio is 1, it means that the odds of disease is equal in $T$ and $C$, and a value greater than 1 will mean that odds of disease is higher among treatment compared to that in the control. Conceptually, though a proportion is easier, as an index, to understand than the odds, the latter is generally used. If a odds ratio is denoted by OR, then from equation 9.20, we have

$$\log(\text{OR}) = \log(f_T) + \log(f_{C'}) - \log(f_{T'}) - \log(f_C) \quad (9.21)$$

The standard error of log (OR) is obtained as (Pagano and Gauvreau, 2000)

$$\text{SE}(\log(\text{OR})) = \sqrt{\frac{1}{f_T} + \frac{1}{f_{C'}} + \frac{1}{f_{T'}} + \frac{1}{f_C}} \quad (9.22)$$

Therefore, under the null hypothesis that the odds ratio is equal to 1,

$$z = \frac{\log(\text{OR})}{SE(\log(\text{OR}))}$$

follows a normal distribution with 0 mean and standard deviation equal to 1. If the alternative hypothesis is that $z$ is either less than or greater than 1, a value of $z$ more than 1.96 will mean that the chance is 5 per cent or less that the sample belongs to a population having $z$ as equal to 1. In other words, the null hypothesis of equality of the two odds is rejected.

## 9.3 IMPACT WHEN CLUSTERS ARE ALLOCATED RANDOMLY

Analysis of impact will differ if clusters are allocated instead of individual units. It is possible to consider either a cluster sampling where all the units belonging to a selected cluster is included or it can be selected in two or more stages. For understanding the impact of different government programmes in rural India, villages are often considered as suitable clusters. Some villages can be considered for implementation of the programme (treatment) and other villages will form the control (without the programme). For impact analysis, both the possibilities, either entire village or selected individuals from each selected village, could be utilized as the units.

Clusters can be allocated randomly to treatment arms. Allocation can also be done after subdividing the clusters into strata. Since village size is often found to be an important factor influencing its development, it can be effectively used as a covariate for the stratification. The impact analysis, when clusters are used, can be done (Hayes and Moulton, 2009) in two different ways: i) at a cluster level and ii) at an individual level.

Let $a_i$ $(i = 1, 2)$ be the number of clusters selected in $T$ $(i = 1)$ and $C$ $(i = 2)$ and $n_{ij}$ $(j = 1, 2, \ldots, a_i)$, the number of individuals in $(i, j)$th cluster. Also, let

$$n_1 = n_{11} + n_{12} + \cdots + n_{1a_1}$$

and

$$n_2 = n_{21} + n_{22} + \cdots + n_{2a_2}$$

be the total number of individuals on which outputs are available in $T$ and $C$, respectively.

### 9.3.1 ANALYSIS AT CLUSTER LEVEL

There are $a$ $(= a_1 + a_2)$ clusters available to examine how an output variable differ between $T$ and $C$. Let $y_{ijk}$ be the output for $k$th observation in $j$th cluster of $i$th treatment. If $\bar{y}_{ij\cdot}$ is the mean of the $j$th cluster for $i$th treatment, then

$$\bar{y}_{ij} = \frac{\sum_{k=1}^{n_{ij}} y_{ijk}}{n_{ij}}$$

is considered as an observation. The data structure can be represented as follows:

| Treatment | | Control | |
|---|---|---|---|
| S. No. of cluster | Mean | S. No. of cluster | Mean |
| 1 | $\overline{y}_{11}$ | 1 | $\overline{y}_{21}$ |
| 2 | $\overline{y}_{12}$ | 2 | $\overline{y}_{22}$ |
| $J$ | $\overline{y}_{1j}$ | $j$ | $\overline{y}_{2j}$ |
| $a_1$ | $\overline{y}_{1a_1}$ | $a_2$ | $\overline{y}_{2a_2}$ |
| Total | $\overline{y}_1 = \frac{\sum_{j=1}^{a_1} y_{1j}}{a_1}$ | Total | $\overline{y}_2 = \frac{\sum_{j=1}^{a_2} y_{2j}}{a_2}$ |

If $y$ is a continuous variable, a $t$-test, as discussed in Section 9.2, can be employed to establish the significance of the difference between $\overline{y}_1$ and $\overline{y}_2$. For this,

$$S_1^2 = \frac{\sum_{j=1}^{a_1} (\overline{y}_{1j} - \overline{y}_1)^2}{a_1 - 1} \text{ and } S_2^2 = \frac{\sum_{j=1}^{a_2} (\overline{y}_{2j} - \overline{y}_2)^2}{a_2 - 1}.$$

One can also consider the non-parametric tests or the other tests, in case $y$ happens to be dichotomous, as mentioned in Section 9.2. Similarly, as mentioned in Section 9.3, a suitable test can be used if clusters are allocated independently within each stratum. Because cluster is considered as a unit of observation, the clustering effect is not considered. Also, it needs to be mentioned that there is no need to use any weight for obtaining the estimate of the cluster means. This is true, irrespective of whether a cluster sampling or subsampling is used to obtain observations within a cluster. It needs to be mentioned that a cluster is considered as one unit. Since clusters are allocated randomly to $T$ and $C$ and the comparison is essentially between them, the use of weights is not required.

One criticism of this approach is that it ignores variations within a cluster. In fact, the cluster mean is used as a representative unit within a cluster and all the other units are disregarded and hence wasted. However, it provides a robust estimate and particularly useful if the number of clusters in an experiment is not large enough, around 20 or less in a treatment arm (Hayes and Moulton, 2009).

### 9.3.2 Analysis at Individual Level

In this case, mean of $y_{1jk}$ for $T$ considering all $n_1$ and, similarly, mean of $y_{2jk}$ for C based on $n_2$ individuals would be considered for the comparison. The mean and standard error of an estimate is obtained as usual, discussed earlier in Chapter 2 (for cluster sampling) or Chapter 3 (for multistage design). Although $n_1$ and $n_2$ observations, instead of $a_1$ and $a_2$, in the earlier section, are used for the comparison here, the effect of clustering is taken into account as discussed in the earlier chapters. As mentioned in Section 3.3, it would be a ratio estimate. For its sampling variance, the number of clusters should be large enough. Hence, if number of clusters is not large enough (more than 15), it would be better to use the cluster level analysis (Hayes and Moulton, 2009).

Also, in case the designs used to select the samples are not self-weighting, use of appropriate weights has to be employed to minimize the bias. As discussed in Chapter 7, even for a self-weighting design, differential non-response in $T$ and $C$ can introduce bias while comparing the two estimates. Hence, it is advisable to use weights to take care of the effect of non-response. Another point worth remembering is about the size of samples in each cluster. As emphasized in the earlier chapters, less the variation in the cluster size, better would be the reliability of the comparison between the two estimates. Even if a cluster level analysis is used, an inherent assumption in the comparison is about the equality of the two variances. A large variation in the cluster sizes can be detrimental to this assumption. For an individual level analysis, there will be a need to adjust the sampling variance to take into account the design effect. In other words, the assumption of a sampling variance of the difference as $\frac{2s^2}{n}$ (if sample size is $n$ in both $T$ and $C$) will not hold good. It would be necessary to compute the sampling variance of the cluster design that is being used. Alternatively, if the design effect is available, one can modify the variance accordingly (Section 7.4.1.3).

## 9.4 IMPACT WHEN STRATIFICATION IS USED BEFORE ALLOCATION

In this case units are allocated randomly within each stratum. A change in the analysis will be required which has been discussed separately according to whether units or clusters are alloted.

### 9.4.1 When units are allocated

In this case, allotment of units to $T$ and $C$ is done independently within each stratum. One can have two or more strata. In such events, where units are allocated independently and randomly within each stratum, all the tests indicated in Section 9.2 are applicable to test the difference within each stratum.

Let us discuss it considering the example presented in Section 8.3. It was to study the effect of antitoxin in enhancing the survival chance of a patient suffering from a disease. The covariate used for the stratification is the condition of a patient (moderately or severely ill) when the treatment started. It will not be appropriate to merge the data in the two strata and perform one test (e.g., the chi-square test) to assess the impact. The assumption behind getting the expected number in a cell requires (mentioned in Section 9.2.2.1) that the probability of surviving is identical, under the null hypothesis, in $T$ and $C$. This may not be tenable when interaction is present. It would be better to check whether the effect of a treatment varies among those who are moderately ill and separately also among those who are severely ill. A difference in the treatment effect between the two strata will mean the presence of an interaction. If there is no interaction, the two effects can be pooled to get the overall estimate of the effect.

If output variable is measured in a continuous scale, it would be advisable to judge the significance of the difference between the two element variances before combining them with the help of equation 9.4. If they are significantly different, equation 9.5 should better be used instead of equation 9.3.

Naturally, as the number of stratum increases, the task of carrying out the test within each stratum becomes more tedious. Also, there is a need to ascertain that sufficient sample sizes are available for conducting a test separately within each stratum. The number of strata is, therefore, kept as small as feasible. Multivariate analysis, such as multiple regressions or logistic regressions, have been widely used for obtaining the effect of an independent variable on a dependent variable, simultaneously adjusting for the effects of a host of other independent variables (Retherford and Choe, 1993).

#### 9.4.1.1 Two-way ANOVA test

When the dependent variable under study is continuous, a two-way ANOVA test is suitable, particularly if the number of strata is large. It is a two-way partition of the total variation in the $y$ variable. Apart from variations due to treatment as in one-way ANOVA (Section 9.2.1.3), the variation is partitioned into that due to the strata (which is known as block in ANOVA parlance), in addition to the treatment and error. Let $y_{ijk}$ denote the $k$th ($k = 1, 2, ..., n$) observation in $j$th ($j = 1, 2, ..., \mathrm{l}$) block of $i$th ($i = 1, 2, ..., m$) treatment. The data structure can be represented as follows:

| | | Strata | | | |
|---|---|---|---|---|---|
| **Treatment** | **Observations** | **1** | **2** | ***j*** | **l** |
| 1 | 1 | $y_{111}$ | $y_{121}$ | $y_{1j1}$ | $y_{1l1}$ |
| | 2 | $y_{112}$ | $y_{122}$ | $y_{1j2}$ | $y_{1l2}$ |
| | $k$ | $y_{11k}$ | $y_{12k}$ | $y_{1jk}$ | $y_{1lk}$ |
| | $n$ | $y_{11n}$ | $y_{12n}$ | $y_{1jn}$ | $y_{1ln}$ |
| 2 | 1 | $y_{211}$ | $y_{221}$ | $y_{2j1}$ | $y_{2l1}$ |
| | 2 | $y_{212}$ | $y_{222}$ | $y_{2j2}$ | $y_{2l2}$ |
| | $k$ | $y_{21k}$ | $y_{22k}$ | $y_{2jk}$ | $y_{2lk}$ |
| | $n$ | $y_{2ln}$ | $y_{22n}$ | $y_{2jn}$ | $y_{2ln}$ |
| $i$ | 1 | $y_{i11}$ | $y_{i21}$ | $y_{ij1}$ | $y_{il1}$ |
| | 2 | $y_{i12}$ | $y_{i22}$ | $y_{ij2}$ | $y_{il2}$ |
| | $k$ | $y_{i1k}$ | $y_{i2k}$ | $y_{ijk}$ | $y_{ilk}$ |
| | $n$ | $y_{i1n}$ | $y_{i2n}$ | $y_{ijn}$ | $y_{iln}$ |
| $m$ | 1 | $y_{m11}$ | $y_{m21}$ | $y_{mj1}$ | $y_{ml1}$ |
| | 2 | $y_{m12}$ | $y_{m22}$ | $y_{mj2}$ | $y_{ml2}$ |
| | $k$ | $y_{m1k}$ | $y_{m2k}$ | $y_{mjk}$ | $y_{mlk}$ |
| | $n$ | $y_{m1n}$ | $y_{m2n}$ | $y_{mjn}$ | $y_{mln}$ |

TSS can be obtained as

$$\text{TSS} = \sum_{i=1}^{m} \sum_{j=1}^{l} \sum_{k=1}^{n} (y_{ijk} - \bar{y})^2,$$

where

$$\bar{y} = \frac{\sum_1^m \sum_1^l \sum_1^n y_{ijk}}{mln}$$

is the overall mean. Alternatively, TSS, showing the total variation in the individual observations can be written as

$$\text{TSS} = [(y_{111}^2 + y_{112}^2 + \cdots y_{1ln}^2) + (y_{211}^2) + (y_{211}^2 + y_{212}^2 + \cdots y_{2ln}^2) + \cdots (y_{m11}^2 + y_{m12}^2 + \cdots y_{mln}^2] - \frac{G^2}{mln} \tag{9.23}$$

$G$ is the sum total of all the $y$ values, that is,

$$G = \sum_{i=1}^{m} \sum_{j=1}^{l} \sum_{k=1}^{n} y_{ijk}$$

The treatment sum of square (Tr.SS) can be obtained as

$$\text{Tr.SS} = \frac{y_1^2 + y_2^2 + \cdots + y_i^2 + \cdots + y_m^2}{ln} - \frac{G^2}{mln} \tag{9.24}$$

where $y_i$ is the sum of all $1 \times n$ units in $i$th treatment, that is,

$$y_i = \sum_{j=1}^{l} \sum_{k=1}^{n} y_{ijk}$$

The treatment sum of square explains the part of the total variations that is due to the treatment. For example, if $m = 2$, it will show the variations that is explained by a difference between $T$ and $C$. Larger the value of the treatment sum of squares, higher will be the chance that the treatment effect is different from that of the control.

The block sum of squares (BSS), showing how $y$ values within the strata vary from each other, is measured as

$$\text{BSS} = \frac{y_1^2 + y_2^2 + \cdots + y_j^2 + \cdots + y_l^2}{mn} - \frac{G^2}{mln} \tag{9.25}$$

It should be noted that $y_j$ is the sum of $y$ values in $j$th block (strata) for $m$ treatments each having $n$ observations. In other words,

$$y_j = \sum_{i=1}^{m} \sum_{k=1}^{n} y_{ijk}$$

The total variation in $y$ values (TSS) is expected to be equal to the variations explained by the treatments (Tr.SS) and the blocks (BSS). Any unexplained variations that still remains, designated as the error sum of squares (ESS), is estimated as

$$\text{ESS} = \text{TSS} - \text{Tr.SS} - \text{BSS} \tag{9.26}$$

Under the null hypothesis that the treatment means in the population $(\bar{Y}_i)$ are equal,

$$H_0 = \bar{Y}_1 = \bar{Y}_2 = \cdots = \bar{Y}_m$$

the ratio of the two sum of squares

$$F = \frac{\dfrac{\text{Tr.SS}}{(m-1)}}{\dfrac{\text{ESS}}{(mln - m - l + 1)}} \tag{9.27}$$

follows an $F$-distribution with $(m-1)$ and $(mln - m - l + 1)$ degrees of freedom. As mentioned in the case of one-way ANOVA, one can calculate $F$ and compare it with its tabulated value to decide whether the null hypothesis is to be rejected or not.

For testing the effect of block (stratification), one can similarly test the BSS by calculating

$$F = \frac{\dfrac{\text{BSS}}{(l-1)}}{\dfrac{\text{ESS}}{(mln - m - l + 1)}} \tag{9.28}$$

### Example 9.3

Considering the data provided in Example 9.2, examine whether the level of education of a female has any role in influencing the impact of the incentive scheme in augmenting marriage age among the females.

In this case, we consider the level of education as a block. The value of TSS and Tr.SS will remain same as obtained in Example 9.2.

$$\text{TSS} = 630.60; \text{Tr.SS} = 131.01.$$

Also,

$$\frac{G^2}{60} = 20063.4$$

$B_1$ = sum of ages at marriage among women who were educated less than secondary

= 488.93

$B_2$ = sum of ages at marriage among women educated at least secondary

= 608.25

$$\text{BSS} = \frac{488.93^2}{30} + \frac{608.25^2}{30} - 20063.4 = 237.29$$

Therefore,

$$\text{ESS} = 630.60 - 131.01 - 237.29 = 262.30.$$

The degree of freedom for the ESS will be 57 (= 59 – 1 – 1). To test the significance of the BSS from equation 9.28,

$$F = \frac{\frac{\text{BSS}}{(l-1)}}{\frac{\text{ESS}}{(60-2-2+1)}} = 51.56$$

at 1 and 57 degrees of freedom. This is highly significant which means that the treatment effect varies according to the level of education.

### 9.4.2 When clusters are allocated

In this situation, the impact estimation can be done within each stratum. The analysis can be done either at the cluster level or at the individual level (Section 9.3). In the former, the cluster level summary indicators can be obtained and an appropriate test indicated in Section 9.3.1 can be used. Since cluster level indicators are used, there would be no need to take into account the clustering effect. Usually, the number of clusters per stratum would be small. There are exact tests available which may become useful in such cases (discussed in Section 9.5.2.1).

## 9.5 IMPACT IN PAIR MATCHING

Suppose, two sets of observations on $T$ and $C$, which are pair matched, are denoted as $y_{Tk}$ and $y_{Ck}$ ($k = 1, 2, ..., n$), that is, $y_{Tk}$ and $y_{Ck}$ denote the outputs in $k$th pair for units belonging to $T$ and $C$, respectively. Since a restriction is imposed that the two units in a pair need to be alike in terms of the disturbing variables, it is apparent that the $2n$ observations are not drawn independently. Generally, one of them, $y_{Tk}$ is drawn randomly and $y_{Ck}$ is selected as the matched group. Therefore, only $n$ values of $d_k$ can be considered to be randomly distributed and analysed, as shown in the data structure given below. We have

$$d_k = y_{Tk} - y_{Ck}$$

as $n$ observations instead of $2n$ observations, which are considered to be allocated randomly.

When cluster is the unit of observation, pair matching is rare because of the difficulty in matching them. We discuss the estimation when individuals are paired, in the table below. The $2n$ observations in a pair-matched experiment can be arranged as follows:

| Pair | $T$ | $C$ | Difference ($d$) |
|---|---|---|---|
| 1 | $y_{T1}$ | $y_{C1}$ | $d_1$ |
| 2 | $y_{T2}$ | $y_{C2}$ | $d_2$ |
| : | : | : | |

| Pair | T | C | Difference (*d*) |
|---|---|---|---|
| $k$ | $y_{Tk}$ | $y_{Ck}$ | $d_k$ |
| : | : | : | |
| $n$ | $y_{Tn}$ | $y_{Cn}$ | $d_n$ |

### 9.5.1 Variables Measured in Interval Scale

When the output variable is continuous, a *t*-test under the null hypothesis of equal population mean of the outputs in *T* and *C*, can be written as

$$t = \frac{\bar{d}}{\frac{s_d}{\sqrt{n}}} \tag{9.29}$$

Where $\bar{d}$ is the mean of the *d* values and $s_d$ its standard deviation calculated as

$$s_d = \sqrt{\frac{\sum_{k=1}^{n}(d_k - \bar{d})^2}{n-1}}$$

The value of *t* can be tested against *t*-statistic with $(n - 1)$ degrees of freedom.

### 9.5.2 Dichotomous Variable

In this case, the outputs take two values, whether a characteristic is present (=1) or not (= 0). It is easy to see that $d_k$ can have the following four categories:

**(i)** equal to 1 ($y_{Tk} = 1$ and $y_{Ck} = 0$),
**(ii)** equal to 0 ($y_{Tk} = 1$ and $y_{Ck} = 1$),
**(iii)** equal to 0 ($y_{Tk} = 0$ and $y_{Ck} = 0$),
**(iv)** equal to –1 ($y_{Tk} = 0$ and $y_{Ck} = 1$).

In other words, the *n* values of $d_k$ can be arranged in a $2 \times 2$ contingency table as follows:

| Characteristic of the output in *T* | Characteristic of the output in C | | |
|---|---|---|---|
| | Present | Absent | Total |
| Present | $y_{11}$ | $y_{10}$ | $y_{11} + y_{10}$ |
| Absent | $y_{01}$ | $y_{00}$ | $y_{01} + y_{00}$ |
| Total | $y_{11} + y_{01}$ | $y_{10} + y_{00}$ | $n$ |

Let $p_T$ and $p_C$ denote the proportion of the cases having the attribute in *T* and *C*, respectively, and $y_{11}$, for example, represents the number of cases where the characteristic is present in both *T* and *C*.

$$p_T = \frac{y_{11} + y_{10}}{n}$$

$$p_C = \frac{y_{11} + y_{01}}{n}$$

One needs to check if $p_T$ is higher (statistically) compared to $p_C$. Since the factor $y_{11}$ is common in both,

$$p_T - p_C = \frac{y_{10} - y_{01}}{n} \tag{9.30}$$

The standard error of $(p_T - p_C)$ is given by McNemar (1947) as

$$\text{Standard error of } (p_T - p_C) = \frac{\sqrt{y_{10} + y_{01}}}{n} \tag{9.31}$$

To test whether $p_T$ is significantly different from $p_C$, one can also calculate

$$\chi^2 = \frac{\left(|y_{10} - y_{01}| - 0.5\right)^2}{(y_{10} + y_{01})}$$

It has been shown that it follows a chi-square distribution with 1 degree of freedom (Edwards, 1948).

#### 9.5.2.1 Exact binomial test

If either $y_{10}$ or $y_{01}$ is small and their sum is less than 25, it is better not to employ the above test. Instead, an exact binomial test can be undertaken to test the significance of the difference. Let $b$ denote the total number of the discordant pairs, that is,

$$b = y_{10} + y_{01}$$

The probability of obtaining at least $x$ number of observations belonging to $T$ (where $T$ is present but not C) in a binomial trial (with either $T$ or C) under the null hypothesis of equal probability can be obtained as

$$= \sum_{x=y_{10}}^{b} {}^{b}C_{x} \left(\frac{1}{2}\right)^{x} \left(\frac{1}{2}\right)^{b-x}$$

If this probability is less than or equal to 0.05, the null hypothesis of equality can be rejected. In fact, all the statistical packages provide for such computation for values of $b$ less than 25.

## EXAMPLE 9.4

In a study, 40 pairs of observation on whether a person is given a vaccination ($T$) or not ($C$), suffered from a disease ($y$), are available. The values of $y$ (=1 if suffered and =0, if not), are given below.

| S. No. | 1 | 2 | 3 | 4 | 5 | 6 | 7 | 8 | 9 | 10 |
|---|---|---|---|---|---|---|---|---|---|---|
| $y_T$ | 0 | 0 | 0 | 1 | 0 | 0 | 1 | 0 | 0 | 1 |
| $y_C$ | 0 | 0 | 1 | 1 | 0 | 1 | 1 | 1 | 1 | 0 |
| S. No. | 11 | 12 | 13 | 14 | 15 | 16 | 17 | 18 | 19 | 20 |
| $y_T$ | 1 | 1 | 1 | 0 | 0 | 0 | 0 | 0 | 0 | 0 |
| $y_C$ | 1 | 1 | 1 | 0 | 0 | 1 | 1 | 0 | 0 | 0 |
| S. No. | 21 | 22 | 23 | 24 | 25 | 26 | 27 | 28 | 29 | 30 |
| $y_T$ | 1 | 0 | 0 | 0 | 0 | 1 | 0 | 1 | 0 | 0 |
| $y_C$ | 1 | 1 | 0 | 0 | 0 | 1 | 1 | 1 | 0 | 0 |
| S. No. | 31 | 32 | 33 | 34 | 35 | 36 | 37 | 38 | 39 | 40 |
| $y_T$ | 0 | 1 | 0 | 0 | 0 | 0 | 0 | 0 | 0 | 1 |
| $y_C$ | 0 | 0 | 1 | 0 | 0 | 0 | 1 | 0 | 0 | 0 |

Use the above data to test whether the proportion suffering from the disease vary according to the vaccination status of a person.

The above data can be arranged in a tabular form as

| | Persons among *C* by categories of *y* | |
|---|---|---|
| Persons among *T* by categories of *y* | Suffered | Not suffered |
| Suffered | $y_{11} = 8$ | $y_{10} = 3$ |
| Not suffered | $y_{01} = 10$ | $y_{00} = 19$ |

**McNemar's test**

Under the null hypothesis that the two proportions $p_T$ and $p_C$ are the same (there is no effect of the vaccine), the value of $\chi^2$ can be calculated as

$$\chi^2 = \frac{(7-0.5)^2}{3+10} = 3.25$$

This value 3.25 with 1 degree of freedom is significant and hence the null hypothesis can be rejected. This means that the vaccination is effective in reducing the proportion suffering compared to that among those who were not vaccinated.

**Exact binomial test**

There are altogether 13 (discordant) cases. Each case describes a pair with two events. First refers to the situations where vaccinated persons were found to suffer from the disease but persons without the vaccination did not suffer ($y_{10} = 3$). The other indicates that a vaccinated person did not suffer, whereas a person not vaccinated suffered ($y_{01} = 10$). It can be seen that

$$y_{10} + y_{01} = 13$$

Clearly, higher the value of $y_{01}$ (or lower the value of $y_{10}$), greater is the likelihood that the vaccine is effective in curbing the disease. The observed values of the two are 3 and 10, respectively. For the exact test, one needs to obtain the probability of getting $y_{01}$ to be 10 or more than 10, keeping the total cases as 13. Alternatively, it can be the probability of getting $y_{10}$ to be 3 or less than 3. Under the null hypothesis, the two probabilities of getting a pair of either first or the second type are equal. Hence, the required probability is

$$13_{C_{10}}\left(\frac{1}{2}\right)^3\left(\frac{1}{2}\right)^{10} + 13_{C_{11}}\left(\frac{1}{2}\right)^2\left(\frac{1}{2}\right)^{11} + 13_{C_{12}}\left(\frac{1}{2}\right)^1\left(\frac{1}{2}\right)^{12} + 13_{C_{13}}\left(\frac{1}{2}\right)^0\left(\frac{1}{2}\right)^{13}$$

$$= (286+78+13+1)\times\left(\frac{1}{2}\right)^{13} = 378\times 0.00012 = 0.045$$

Since this probability is small, less than 0.05, it is unlikely to have occurred if the null hypothesis was true. The null hypothesis is, therefore, rejected. It can be said that the treatment is effective, those with the vaccine is less likely to suffer from the disease compared to persons without a vaccination.

### 9.5.3 Non-Parametric Test

Non parametric tests can be employed to measure statistical significance of the difference between the outputs in $T$ and $C$.

**(a) Wilcoxon signed rank test**

In the Wilcoxon rank sum test for the difference between two independent samples means all the $2n$ sample values are considered for the ranking. For pair matching, $n$ values of $d_k$ are ranked from the lowest to the highest one. This is true, irrespective of whether the matching is done with the help of propensity score matching or otherwise.

While comparing $y_{Tk}$ with $y_{Ck}$, in calculating $d_k$, it is natural to expect that some values would be positive indicating a higher value of $y_{Tk}$ compared to $y_{Ck}$ and others would be negative denoting the opposite situation. Given a situation where one expects the treatment to have a positive impact, higher the positive ranks, greater the possibility that the treatment has an impact.

For the purpose of ranking, absolute values of $d_k$ are considered and question of ties is resolved as usual by taking the average value. Also, the cases where $y_{Tk}$ equals $y_{Ck}$ are deleted. Assuming that $n$ is the number of pairs available for carrying out the test (after deletion), a similar rank sum test, as employed in the earlier case, can be applied. Suppose,

$\text{Sum}_T$ = aggregates of the ranks of $d_k$ values where $y_{Tk}$ is greater than $y_{Ck}$
$\text{Sum}_C$ = aggregates of the ranks of $d_k$ values where $y_{Tk}$ is less than $y_{Ck}$
than $y_{Ck}$

If $n$ denotes the total number of pairs available for the test, that is,

$$\text{Sum}_T + \text{Sum}_C = \frac{n\ (n+1)}{2}$$

then smaller of the two aggregates can be checked with the corresponding tabulated values to infer whether the gap between the two aggregates is large enough to justify the statistical significance of a treatment effect. The tabulated values, both for applying the rank sum test in the case of two independent samples and that for paired data, are readily available (Wilcoxon and Wilcox, 1964).

**(b) Normal approximation**

For large sample size (greater than or equal to 25), one can use the normal approximation as indicated in the case of the two independent samples. The $z$-statistic, for the two-tailed test under the null hypothesis that the population means of $T$ and $C$ are equal, can be performed with $\text{Sum}_C$ (assuming that it is less than $\text{Sum}_T$, otherwise with the latter). With

$$E\ (\text{Sum}_C) = \frac{n(n+1)}{4} \tag{9.32}$$

and

$$V\ (\text{Sum}_C) = \frac{n\ (n+1)\ (2n+1)}{24} \tag{9.33},$$

$z$ can be obtained as

$$Z = \frac{(\text{Sum}_C - E\ (\text{Sum}_C))}{\sqrt{V\ (\text{Sum}_C)}} \tag{9.34}$$

## 9.6 MODEL-BASED ANALYSIS

An alternative way, besides a design, could be to employ a model-based approach to measure the impact (Chapter 6). In fact, the use of regression analysis to explore relationship between variables is widespread. Its use is mainly because it does not require the act of explicitly balancing the distribution of disturbing variables among $T$ and $C$. It makes possible to explore the relationship between variables under variety of different situations, often with a very little cost, for example, from available large-scale sample survey data.

It is important, however, to note that a model-based analysis requires that a data being analysed follow certain assumptions. Unless the assumptions, at least the important ones are met by the data, the technique employed is not likely to provide reliable estimate of a treatment effect. Such analysis is described well in most textbooks of statistics.

Since the focus of the present book is on designs, only a brief discussion on model-based approaches will be presented. Some idea about how to build a model, derivation of the effect of a variable in a model and its interpretation is helpful in designing an evaluation study. In spite of utmost care taken in a design, it is not always possible to exercise all the necessary controls. The application of model-based approaches might be of help in restoring some of the balance.

### 9.6.1 Multiple Regression Analysis

In its simplest form, the statistical relationship between a dependent variable $y$, measured in an interval scale, with independent variables ($x_i$s) can be depicted by a linear model as

$$y = a + b_1 x_1 + b_2 x_2 + \cdots + b_p x_p + \varepsilon \qquad (9.35)$$

It means that for given values of the independent variables, if the constants $a$ and $b_i$ are known, it is possible to get a corresponding estimate of $y$. If we have a set of observations on $y$, $x_1$, $x_2$, ..., $x_p$, it will be possible to compare the observed values of $y$ with that of their estimates based on equation 9.35. The difference between an actual value of $y(y_i)$ and its estimate is termed as error ($\varepsilon_i$). These errors are minimized while estimating the constants.

A regression coefficient, $b_i$, measures the relationship between $x_i$ with $y$. The relationship or the effect is supposed to be a causal effect. That is, the effect of $x_i$ will show the extent to which $y$ changes for a unit change in $x_i$ keeping constant all the other independent variables present in the model. Hence, if $x_i$ denotes a treatment (two groups $T$ and $C$ denoted by 1 and 0) and the other independent variables are the disturbing variables, the estimated value of $b_i$ will be the treatment effect.

The regression coefficient $b_i$ will have a sampling variance, which can be estimated easily. The standard error of $b_i$ is given by

$$\text{Standard error of } (b_i) = \frac{b_i}{\sqrt{n}} \qquad (9.36)$$

where $n$ is the number of observations available for estimating the coefficients. It needs to be checked whether $b_i$ is significantly different from zero. For example, if $b_i$ is greater than twice (~1.96) its standard error, it can be regarded as having an effect on $y$.

In the case of an evaluation design, as discussed in the previous chapters, covariates are usually controlled at the time of allocation. Hence, the statistical adjustment through a regression equation is not usually needed. If required, a regression equation between $x_1$ (treatment) and the output variable $y$ measured in an interval scale can be written as

$$y = a + b_1 x_1$$

For example, $y$ might denote hours of sleep and $x_1$ the treatment which takes a value of 1 for those who are administered a sleeping pill whose effect is being measured and 0 for those in the control group. Also, if there are $n$ observations on both treatment and control, one will have $2n$ observations on $y$ and $x_1$ ($n$ of these will have $x_1 = 1$ and remaining with $x_1 = 0$). The data set can be used to calculate the values of $a$ and $b_1$. A statistical package, such as SPSS or STATA, can also be used to estimate the parameters. The packages also provide standard errors of the estimates.

It can be easily checked that $b_1$ is equal to a difference in the mean values of $y$ in treatment and control groups. In this case, there are no covariates adjusted in the model because they are already assumed to have been controlled in the design.

A covariate, which, for some reasons, could not be balanced in a design, can also be included, if necessary, in the regression. If age ($x_2$) is such a variable, the model can be changed as

$$y = a' + b_1' \; x_1 + b_2' \; x_2$$

The estimated value of $b_1'$ will provide the effect of $x_1$ after controlling the effect of $x_2$.

It might be of interest to examine whether the treatment ($x_1$) interacts with any other covariates (say $x_2$). To accomplish this, a product term $x_3$, where

$$x_3 = x_1 x_2$$

can be introduced into the model, that is, it can be written as

$$y = a' + b_1' \; x_1 + b_2' \; x_2 + b_3' x_3$$

The estimated value of the coefficient $b_3'$ will assist in understanding about the interaction. If it is found to be statistically significant, it would mean a change in the effect of $x_1$ according to the levels of $x_2$. If $x_2$ is measured in interval scale, then it can be interpreted that the effect of the treatment ($x_1$) changes with a unit change in $x_2$.

The independent variable $x_i$ can be measured in an interval scale, or in an ordered or in nominal scale. For example, if age is an independent variable, it can be in interval scale from 0, 1 and so on for any value beyond 100 years. If age is depicted in, say, three categories as 0–17, 18–64 and 65 and above, it would be in order scale because there is a hierarchy; the individuals in the second category are older than the previous one but younger than the next one. A variable with three categories can be represented by two dichotomous variables, known as dummy variables (Fleiss, 1973). Some variables can be nominal with two or more categories. A three-category nominal variable can be, say, migration status of persons, as a resident born in an area, migrant from a neighbouring area and migrant from distant area. Generally, there is no hierarchy in a nominal variable. A variable with $n$ categories is represented by $(n - 1)$ dummies.

#### 9.6.1.1 Modifications in the case of cluster sampling

If cluster sampling is used to obtain the observations on $T$ and $C$, as mentioned in Chapters 2 and 3, a modification would be required to take into account the clustering effect. The effective sample size reduces because of the clustering effect. The sampling variance of both $T$ and $C$ will increase, and so also their difference. The sampling variance of a regression coefficient of the treatment is likely to be higher than that mentioned in equation 9.36. One needs information on design effect of a regression coefficient to modify its variance. Although the design effect for estimating the sampling variance of mean can be computed (as a ratio of the variance of the design and the variance in the case of simple random sampling of same size, see equation 3.11), the design effect of a regression coefficient is not known and difficult to estimate.

It has been, however, shown that the design effect of a regression coefficient will be lower than that of the mean (Kish, 1987). One can, therefore, assume that the design effect for the regression coefficient will be equal to that of the mean to modify the variance of the former. This revised estimate can be used to test the null hypothesis for a null value of the regression coefficient.

Another alternative approach to the analysis, which is presently recommended, is to use the statistical packages, such as GEE or MLE, that automatically takes into account the clustering effect.

### 9.6.2 Logistic Regression

When a dependent variable $y$ is dichotomous, an appropriate model to use is the logistic regression. Since $y$ takes only two values 0 and 1, expressing it as a linear function of a set of independent variables often fails to provide a meaningful explanation of its variability. A more appropriate way to set up the regression is to consider

$$\text{logit } p = \log \frac{p}{1-p}$$

as a dependent variable. The parameter, $p$ denotes the proportion of cases with a value of $y$ equal to 1, and $\frac{p}{1-p}$ is the odds of the event, that is, $y$ taking a value of 1 compared to that of zero. The logistic regression between the dependent variable and the set of independent variables ($x_i$) is considered as

$$\text{logit } p = a + b_1x_1 + b_2x_2 + \cdots + b_p\, x_p + \epsilon_i.$$

The only change here is that the dependent variable, $y$ is dichotomous. The impact or the relationship of an independent variable $x_i$ with the dependent variable (logit $p$) can be understood in the same manner. For example, $b_i$ reveals the impact or a change in the dependent variable for a unit increase in $x_i$. The nature of the dependent variable, which is logit $p$, makes its interpretation little complicated. Let us consider an example to further clarify this.

Suppose, the interest is on understanding the impact of a vaccination on curbing the number of individuals who will suffer from a disease, as mentioned in the following data:

| Whether suffered from the disease? | Whether received vaccination? | | |
|---|---|---|---|
| | Yes | No | Total |
| Yes | $N_{11}$ | $N_{21}$ | $N_1$ |
| No | $N_{12}$ | $N_{22}$ | $N_2$ |
| Total | $N_1$ | $N_2$ | N |

Clearly,

$p_1$ = proportion who suffered from the disease among those given the vaccination

$$= \frac{N_{11}}{N_1}$$

$p_0$ = proportion who suffered from the disease among those not given the vaccination

$$= \frac{N_{21}}{N_2}$$

Therefore, odds of disease among those vaccinated

$$= \frac{p_1}{1-p_1} = \frac{N_{11}}{N_{12}}$$

and

odds of disease among those not given vaccination

$$= \frac{p_0}{1-p_0} = \frac{N_{21}}{N_{22}}$$

If it is assumed that the model is

$$\text{logit}\, p = \text{a} + b_1 x_1$$

where $x_1$ is the treatment, then its effect $b_1$ is the difference in $y$ $(\log \frac{p}{1-p})$ values in treatment and the control groups. That is,

$$b_1 = \log\left(\frac{p_1}{1-p_1}\right) - \log\left(\frac{p_0}{1-p_0}\right) = \log\left(\frac{\frac{p_1}{1-p_1}}{\frac{p_0}{1-p_0}}\right)$$

In other words, $b_1$ is the logarithm of the odds ratio. That is, if $b_1$ is positive, the odds of the disease among the treatment (those vaccinated) is higher compared to that in the control. This also means the proportion suffering from the disease is higher among treatment in comparison to the control. The reverse will be the situation if $b_1$ is negative. The statistical significance of the difference can be gauged by using the standard error of the coefficient. To get a better idea about the extent of the difference, $e^{b_1}$ can be computed (which is also provided in the statistical packages).

$$e^{b_1} = \frac{\frac{p_1}{1-p_1}}{\frac{p_0}{1-p_0}} \quad or \quad \frac{p_1}{1-p_1} = e^{b_1} \frac{p_0}{1-p_0}$$

It can be said that the odds of the disease is higher among those vaccinated by a factor, $e^{b_1}$ compared to that in the control group.

### 9.6.3 Assumptions in Regressions

One of the assumptions in the usual regression is that error is independently distributed, in fact, not only of the independent variables but also of the dependent variable. This can get violated if two independent variables are related. It may be noted that the interaction effect of two independent variables is a different issue from the fact that they are related to each other. In impact estimation, independent variables consist of disturbing variables and a treatment. Suppose, a treatment is related to a disturbing variable, that is, the disturbing variable apart from influencing the output also influences with individual's decision to be in $T$ or $C$.

To elaborate, suppose $x$ denotes the set of disturbing variables and $T$ is the dichotomous variable representing the treatment and $y$ is the output. The regression

$$y = a + b_1 x + b_2 T \tag{9.37}$$

assumes that $T$ is not affected by $x$. However, if $T$ is also affected by $x$, it means there will be another equation as

$$T = a' + b_1' x \tag{9.38}$$

It is difficult, in such a situation, to get a correct estimate of the effect of $T$ (i.e., $b_2$) if equation 9.37 is used without considering the other equation (Maddala, 1977; McGinnis, 1958).

Such situation occurs frequently, particularly in evaluation studies dealing with human beings. The socio-economic characteristics of human beings, often considered as disturbing variables, play a role in affecting their involvement in $T$ or $C$. Introduction of a variable, called 'instrumental' variable in equation 9.38 (which is not present in equation 9.37) can rectify the problem.

To elaborate, let us consider, for an example, a study that was undertaken to assess whether increasing awareness of people, particularly in the lower socio-economic group about the human papillovirus vaccine (HPV), used in connection with sexually transmitted diseases, would enhance its acceptability. Suppose, $y$ denotes the persons who accepted the vaccine and the treatment, $T$ denotes whether a person is aware of it or not.

The disturbing variables $x$ constitute the socio-economic factors. It is true that awareness ($T$) of a person would also depend on their level of education and other socio-economic characteristics and hence equation 9.38 needs to be considered, if observational data available from a survey is used. Clearly, education of a person is a disturbing variable and it will not be balanced. This is because the sample consisting of educated person will likely to have unduly higher proportion of those being aware about HPV ($T$) rather than not aware ($C$).

If a design had been used for the evaluation, the data would have been generated in a manner so that it would be automatically balanced; number of educated persons aware and not aware about the vaccine would be similar.

The study considered awareness of another vaccine used in lung cancer as an instrumental variable (Do and Wong, 2012). That is, it is reasonably believed that while awareness of lung cancer vaccine would be related to awareness of HPV vaccine, it will not affect the latter's acceptability. One can write the two equations as

$$y \text{ (acceptability of HPV vaccine)} = a + b_1 x + b_2 T$$

where $T$ is the awareness of HPV vaccine, and

$$T = a' + b_1' x + b_2' x'$$

The introduction of this instrumental variable, awareness about lung cancer vaccine ($x'$), which is believed to be related to $T$ but not $y$ would allow a better estimation of the impact of $T$ on $y$ ($b_2$). There is another approach, known as propensity score method, that can also be used in similar situations (Guo and Fraser, 2010).

## 9.7 KEY POINTS

- A treatment effect is obtained by comparing two outputs, one where the treatment is present ($T$) and the other where it is absent ($C$).
- Two widely used measures of a treatment effect are a difference and a ratio of $T$ and $C$. In the first case, mean or proportion of the two outputs are the commonly used ones. In the second, the odds ratio is generally preferred because of its statistical properties.
- The estimated impact needs a statistical test to understand whether it is real or has arisen due to a sampling fluctuation.
- If units are allocated randomly, the usual statistical tests, such as $t$-test, ANOVA test, etc. can be used to test the significance of the impact. The non-parametric tests, such as Wilcoxon rank sum test or Kruskal–Wallis test can also be employed.
- In the case of stratification along with random allocation within each stratum, an additional check for interaction is required before deciding whether to pool the entire data and analyse or conduct the analysis separately within each stratum.
- In the case of pair matching, the usual $t$-test, McNemar's test or exact binomial test can be employed to test the significance of the impact. The non-parametric Wilcoxon signed rank test is also commonly used for the purpose.

# Chapter 10

# Case Studies

## 10.1 INTRODUCTION

This chapter presents case studies on both survey and evaluation designs. Two case studies, one on a sampling design, namely, that for the NFHS in India and another on an evaluation design conducted in Lucknow, India, in which the authors were involved, have been discussed in greater detail. The purpose has been to demonstrate the steps involved in implementing the designs in the field. In addition, sampling designs for a few other large-scale surveys have also been summarized. Similarly, one application for each of the different evaluation designs has been included. The studies are presented in two separate parts.

## PART I: SAMPLE SURVEY DESIGNS

## 10.2 NATIONAL FAMILY HEALTH SURVEYS, INDIA (NFHS, INDIA)

NFHS are the Indian editions of the DHS conducted in over 70 countries around the world. Three rounds of NFHS have already been conducted in the country: 1992–93, 1998–99 and 2005–06. The coverage and parameters estimated in the three rounds of the survey have been summarized in Table 10.1.

## 10.3 SAMPLING DESIGN OF NFHS

The overall structure of NFHS design is similar to that adopted in other such surveys globally. A household is considered as the basic unit for sample selection. Since there was no viable frame available for selection of households directly, it was necessary to select a sample in multiple stages.

### 10.3.1 Sample size

A general approach was to have 4000 completed interviews with ever married women in a state with a population of 25 million or more, 3000 for those having a population between 2 and 25 million, and 1500 for states with population of 2 million or less. This is because a

TABLE 10.1 A few salient information of the three rounds of NFHS.

| Parameters | NFHS-1 | NFHS-2 | NFHS-3 |
|---|---|---|---|
| Year conducted | 1991–92 | 1998–99 | 2005–06 |
| Coverage (in terms of number of interviewed respondents) | 89,777 ever married women aged 15–49 years from 88,562 selected households | 89,199 ever married women aged 15–49 years from 91,196 selected households | 124,385 married and unmarried women aged 15–49 years and 74,369 men aged 15–54 years, selected from 109,041 selected households |
| Levels at which estimates were required | All India and all the states except Sikkim | All India and all the existing states in 1998–99 | All India, all the existing states and in 8 cities (separately for slum and non-slum areas) |
| Parameters estimated | Usual fertility, family planning and maternal and child health care related indicators | Apart from all the NFHS-1 parameters, women's autonomy, domestic violence, reproductive health and nutritional status of women and their children, including blood samples for anaemia | Apart from all the NFHS-2 parameters, blood samples were collected for the estimation of HIV prevalence in all India and few other states |

*Note: The fourth round of the survey (NFHS-4) has started during 2014. This round of survey intends to provide estimates for a large number of indicators for the 640 districts covering more than 500,000 households in the country.*

sample of 1500 households was assumed to be adequate in terms of providing estimates for most of the demographic and maternal and child health parameters being studied.

Allocation of sample size for each state was not in proportion to their population. The slightly higher sample size for the larger states was mainly to facilitate understanding the socio-economic differentials in a parameter. Having larger sample size for the bigger states also made sense because it reduces, to some extent, the banal effect of the disproportional allocation of sample while obtaining the national level estimates.

The estimated sample size was, however, revised whenever necessary. For example, HIV/ AIDS was estimated in NFHS-3, and it required a larger sample size than was considered in the earlier rounds. The estimate was needed in a state where its prevalence was likely to be around 1 per cent. The required sample size for a state was obtained as (using equation 2.21)

$$n = \frac{1}{(0.14)^2} \times \frac{0.99}{0.01} = 5051$$

This implies that a parameter assumed to be 1 per cent in a population can be estimated with a standard error given by

$$\text{Standard error of } p = 0.14 \times 0.01 = 0.0014$$

Assuming a design effect of 1.25, the total sample size was increased to 6,400 for a state for which estimate of HIV prevalence was needed. Therefore, there is a 95 per cent chance that the parameter will lie in between 0.73 and 1.27 per cent.

### 10.3.2 Choice of PSU

Perhaps the most important aspect of a multistage design is to decide the PSUs. In rural areas, villages are the smallest areal units that have clearly defined boundaries. It thus helps ensure that each and every household/individual in rural areas of a state gets a chance of being selected in the sample. It also provides an excellent opportunity to ensure that the selected sample is well spread. The average size of a village in India is around 300 households. A state has a large number of villages that helps in considering large number of PSUs. This would be beneficial in many ways, such as:

1. It would reduce the clustering effect in a cluster sampling by having large number of independent selections.
2. Larger the number of PSUs selected, smaller is the likely value of $Meff_w$ (equation 3.25) and hence a likely reduction in the sampling variance of an estimate and, finally.
3. It would help to diversify the stratification of PSUs.

The decision about optimum size of a PSU depends on a number of factors. While small size PSUs, as indicated above, are beneficial in many ways, as the size reduces, the possibility of an increase in the value of the intra-cluster correlation ($\rho$) becomes stronger. And as $\rho$ increases, the effective sample size would decrease (equation 3.13). The cost of a survey and the required precision of an estimate would have to be taken into account in deciding the size of PSUs to be considered.

Knowledge of the extent of clustering and the design effect will facilitate the decision regarding what should be a reasonable sample to be taken in PSUs and hence the number of PSUs to have, for a given sample size. Although the design effect is generally available in large-scale surveys, they should be used carefully to obtain idea about the clustering effect (discussed in Section 3.2.2.3).

### 10.3.3 Design for rural area

Generally, a two-stage design had been adopted for the selection of households in all three NFHS. Villages considered as the PSUs were selected using the frame provided by a recent census. Before the selection, the frame was modified to take into account the following two aspects:

(i) merging of small villages and
(ii) stratification

#### 10.3.3.1 Merging of small villages

There are villages in the frame comprising a small number of households. Keeping such villages in the frame can cause inconvenience. At the same time, removing them from the frame leads to violation of the basic tenet (that each and every unit, in a population, should have a non-zero chance of selection) of a probability sampling. To balance the two, very small villages whose cumulative sum forms a negligible proportion of the population, such as less than 5 per cent, were excluded from the frame. Let us suppose that all villages of size less than 10 households were removed from the frame.

Since the decision was to select on an average 30 households in a selected village, its size should be at least 30 or, preferably, greater than 30. This is because smaller villages are often found to have negative rate of growth. Suppose, it is decided to keep the minimum PSU size as 60 households. Therefore, all villages having a population between 10 and 59 would be combined with an adjacent village. The procedure followed for linking small villages is discussed in the following.

For linking, villages in a frame are first arranged according to their geographical location. Suppose,. $X_i$ denotes the size of the $i$th village. Screen the value of $X_i$ serially starting from the top. If $X_i$ is greater than 60, the cut-off decided by the surveyors, continue with the screening. If $X_i$ is less than 60, check the value of $X_{i+1}$. If $X_{i+1}$ is greater than 60, link $X_i$ with $X_{i-1}$, and continue. If $X_{i+1}$ is also less than 60, join the two, $X_i$ and $X_{i+1}$ and check its sum. If the sum is greater than 60, accept it as a PSU and proceed. If their sum is still less than 60, check $X_{i+2}$. If $X_{i+2}$ is greater than 60, then join the two ($X_i$ and $X_{i+1}$) with $X_{i-1}$, and continue. If $X_{i+2}$ is again a small village, join all the three and check their sum. If it is greater than 60 accept it as a PSU, otherwise proceed the same approach by checking $X_{i+3}$. A small example of the procedure is given in Table 10.2.

**Table 10.2** Distribution of villages in a frame consisting of 11 villages before and after the linking procedure

| S. No. of villages | Frame (villages of size ($X_i$)) | S. No. of PSUs (linked villages) | Frame |
|---|---|---|---|
| 1 | 100 | 1 | 100 |
| 2 | 70 | 2 | - 70 |
| 3 | 30 | 3 | 30 + 22 + 10 = 62 |
| 4 | 22 | 4 | 90 |
| 5 | 10 | 5 | 80 + 15 + 20 = 115 |
| 6 | 90 | 6 | 65 + 18 = 83 |
| 7 | 80 | | |
| 8 | 15 | | |
| 9 | 20 | | |
| 10 | 65 | | |
| 11 | 18 | | |

A PSU can, therefore, have more than one village combined, and, if so, they will be treated as one unit for the PSU selection and, if selected, for the household selection as well.

#### 10.3.3.2 Stratification

Stratification of PSUs had been done to the maximum possible extent. Apart from the variables that could be considered for the stratification, its extent depended on the number of PSUs to be selected in a state, which was obtained by dividing the rural sample size by 30 (the average number of households to be selected in a PSU). In NFHS-3, apart from geographic region, following five variables had been considered for the stratification in different states (IIPS and Macro International, 2007): i) village size; ii) female literacy; iii) per cent of SC/ST population; iv) per cent of males in non-agricultural work and v) HIV prevalence (available only at the district level). Among these, village size was used in all the states.

Except in Mizoram, female literacy was used for the stratification in all other states. The variation in female literacy being negligible in Mizoram, it was not worth considering it as a stratification variable. HIV prevalence was used in the states of Andhra Pradesh, Karnataka, Maharashtra, Manipur, Nagaland and Tamil Nadu.

Female literacy was considered to be an important variable that influences fertility as well as maternal and child health in a population. This variable was therefore used for ranking the PSUs within a stratum before their selection. That is, after stratifying the PSUs, in a state, by the other variables, such as geographic subdivision, household size, etc., they were ranked within a given stratum by their level of female literacy. The ranking in a stratum was done either in increasing or decreasing manner and it alternated in successive stratum. That is, if in a stratum the ranking was in an increasing order, in the next it would be in a decreasing manner (for further illustration, see Roy and Pandey, 2008). In four states, namely, Chhatisgarh, Jharkhand, Mizoram and Orissa, percentage of SC/ST population was used for the ranking, instead of female literacy, in NFHS-3 for similar reasons explained above.

It was decided to select between three and five PSUs in a stratum. This along with the total number of PSUs available for the estimation in an area helped in choosing the number of strata. Another criterion followed was to ensure that the size of each stratum (total number of households) did not vary widely (see Section 2.3.5.3), a criterion that the size of any stratum should lie in between $\frac{1}{2}B$ and $2B$, where $B$ is the average size of the strata.

Allocation of a sample, in a state, to its strata, in a domain was close to a proportional allocation. This was also true, in general, when allocation was done for its different domains. However, disproportionate allocation had to be considered for domain wise allocation to meet two objectives. One, if estimates were required to be done for a relatively smaller area. For example, in Himachal Pradesh the sample size had to be jacked up to enable having its urban estimates. Similarly, the size had to be increased for Andhra Pradesh, Delhi, Madhya Pradesh, Maharashtra, Tamil Nadu, Uttar Pradesh and West Bengal where separate estimates were needed for a few of their cities. The other reason for the disproportionate allocation was to try and ensure enough number of HIV cases. To meet this objective, the frame was subdivided into groups according to their HIV prevalence (considered at a district level). Oversampling was done from the group showing high proportion of the prevalence.

While making estimation, the disproportionate allocation of samples had to be compensated by appropriate weights. This will nevertheless increase the coefficient of variation of weights and consequently the value of $\text{Meff}_w$ (equations 2.6 and 2.7) to increase the sampling variance. To what extent it will augment the co-efficient of variation (CV) depends on the degree of departure from the proportionality. For example, for HIV cases, the idea behind the stratification was to have more of such cases and hence allocate more sample size to the regions having higher proportion of the cases. If such regions (having higher proportion of HIV cases) also had higher proportion of persons in their population, the departure from proportionality will be less. The extent of non-response can also contribute to the increase in CV. For illustration, the CV of weights have been computed and presented for a few selected states in Table 10.3.

**Table 10.3** Distribution of CV of weights for households and $Meff_w$ for selected states/country, NFHS-3.

| State/Country | CV of weights | $Meff_w$ |
|---|---|---|
| Himachal Pradesh | 0.4403 | 1.1939 |
| Kerala | 0.0028 | 1.0000 |
| Maharashtra | 0.7299 | 1.5327 |
| India | 0.8515 | 1.7171 |

In Kerala, proportional allocation was used to distribute the sample to its two domains, rural and urban. As a consequence, the CV of weights was negligible. There was no increase in the sampling variance on this count. In Himachal Pradesh, the sample size in urban area had to be increased to provide estimates for it. The sampling variance of an estimate, as a result increases by about 20 per cent. The extent of disproportional allocation was higher in Maharashtra. The highest value of $Meff_w$ was obtained for the country. In spite of the fact that the sample size was very large for the country, the sampling variance of its estimate becomes almost double due to the disproportional allocation of the sample. For example, while the state of Goa with 0.1 per cent of the total population of the country received 3 per cent of the total sample households for the country, the state of Uttar Pradesh with 16 per cent of the country's population, the sample allocation was to the extent of only 9 per cent.

#### 10.3.3.3 Selection of sampling units

The households have been selected in two stages. The process begins with the selection of PSUs by PPS sampling. After arranging the PSUs, required number of them is selected systematically with the help of a single random number using PPS sampling (Section 2.5.1).

Systematic sampling for selection of PSUs should be avoided if there is a fear of having a periodicity that could result in selecting PSUs of similar characteristics (Section 2.4.4). However, steps were taken in NFHS to check the occurrence of such bias. The size measures used at different unit levels were not exactly the same. Also, the arrangement of PSUs in ascending or descending order in alternative strata helped in disturbing the possibility of such oscillatory effects. For example, for a characteristic, such as village size, the distributions of villages selected in the sample and in the population were comparable.

All the households existing in a PSU (consisting of one or more linked villages) were listed through house-to-house visit for the listing. The listing of PSUs was necessary because PSU selection was based not on the current population size but the estimated size as obtained in the recent census. The exact number of households to be selected from a PSU was decided by calculating $f_2$, the probability of selection of a household (Section 3.3.4.2.1).

The value of $f$, the overall sampling fraction, was calculated, in advance, by obtaining the projected population for a state at the survey time. This was done by extrapolating the population available at the recent census at exponential rate of population growth observed at the most recent intercensal period. The estimate was corroborated with the estimated population available from the Expert Committee of Population Projection in the country to arrive at the final estimated value of the population at the survey time. The value of $f_2$, the probability of selection at the second stage, was obtained as

$$f_2 = \frac{f}{f_1}$$

where $f_1$ is the probability of selecting a PSU. The probability of selecting the $i$th PSU is

$$f_1 = \frac{aV_i}{V}$$

$V_i$ is population size (number of households or population) of the $i$th PSU, at the recent census.

The interval of selection of a household in the $i$th PSU ($I$) will be

$$I = \frac{1}{f_2}$$

The number of households expected to be selected, $b_i$, is calculated as

$$b_i = f_2 \times V_i'$$

where $V_i'$ is the size of the $i$th PSU as obtained by the household listing.

The selection of $b_i$ households were done systematically with the help of numbers: $R$, $R+I$, $R+2I$, ..., $R+(b_i-1)I$, where $R$ is random number between 1 and $I$ (Section 2.4.2). The procedure of selection of households in a selected village was, however, slightly different in NFHS-3. Since HIV testing was included in this round, looking at the operational convenience, instead of trying to spread the entire sample of $b_i$ units in $i$th village, a group of five consecutive households were selected. That is, if $b_i$ is 30, six households were chosen systematically. Once a household is selected, four additional households in succession to it were automatically included in the sample.

As discussed earlier (Chapter 3), two issues are important in the context of a multistage design. One is that it should preferably be a self-weighting or an equal probability selection method (EPSEM) design. If that is not feasible, the departure from an EPSEM design should be as small as possible. The departure tends to increase the sampling variance of an estimate by the factor $\text{Meff}_w$. The design, in the present case, within a domain, is self-weighting, assuming no non-response or coverage adjustments.

The other point is about the variations in the sample take ($b_i$). The effort should be to make them equal, as far as possible. It has been suggested (Section 3.3.3) that the coefficient of variation in $b_i$ should not be higher than 2 per cent, preferably less than 1 per cent. There will be a variation in $b_i$ in the present case. Since the value of $b_i$ depends on $f_2$ and $V_i'$, and $f$, $a$ (number of PSUs to be selected) and $V$ are fixed, the variation will depend on differential rate of growth of villages in a state. In other words, $b_i$ will depend on the ratio of $\frac{V_i'}{V_i}$. Theoretically, the ratio should not exhibit large variations. Any error in enumerating either $V_i$ or $V_i'$ will, however, tend to make erratic changes in the ratio. That is, any error in the census results or the listing operation, particularly the former can introduce undue variations. Unexpected change at the time of listing can also occur due to some natural calamities, such as flood or land slide. Also, there can be sudden spurt in construction activities or migration that can influence the rate of growth of some area. Such abrupt changes in the cluster size ($b_i$) can influence the estimation of parameters, particularly the sampling variance (Section 3.3). There is a possibility to reduce the banal effects of such changes. For example, in NFHS, the variations in $b_i$ have been trimmed. Considering the average sample take as 30 households, its value was not allowed to vary beyond 15–60. The design would deviate from an EPSEM due to such adjustment, which was duly incorporated by adjusting the weights. The procedure is illustrated with a small example.

### Example 10.1

A population consist of 2500 households distributed over six PSUs as given below.

| S. No. of PSU | Household size | Cumulative No. of households | Households at the time of listing |
|---|---|---|---|
| 1 | 200 | 200 | |
| 2 | 500 | 700 | |
| 3 | 1000 | 1700 | 1680 |
| 4 | 100 | 1800 | |
| 5 | 400 | 2200 | |
| 6 | 300 | 2500 | 250 |
| Total | 2500 | | 3000 (estimated) |

Forty households are required to be selected by selecting two PSUs and 20 households from each selected PSU. It was decided to restrict the variation in cluster size . The variation is restricted between 16 and 25 households.

It was estimated that the increase in population till the survey time is expected to be by 120 per cent, that is, the estimated population at survey would be 3000 households. Therefore,

$$f = \frac{40}{3000} = 0.013333$$

The interval ($I$) for the selection of PSUs is

$$I = \frac{2500}{2} = 1250$$

The random number $R$ is 1046. The two numbers for the selection are 1046 and 2296, which means PSUs 3 and 6 are selected. The numbers of households listed in the two PSUs are shown in column 4.

**(i) Selection and estimation in PSU 3**

The value of $f_2$ is

$$f_2 = \frac{f}{f_1} = \frac{40}{3000} \times \frac{2500}{2 \times 1000} = \frac{1}{60}$$

Number of households to be selected ($b_i$) will be

$$b_i = \frac{1}{60} \times 1680 = 28$$

Since this is higher than 25 (the upper limit), 25 households will be selected from this PSU. The probability of selection of a household in PSU 3 will therefore be

$$= \frac{2 \times 1000}{2500} \times \frac{25}{1680} = 0.011905$$

Note that it deviates from the self-weighting design; its value is lower than $f$. Had we selected 28 households, it would have been equal to $f$.

**(ii) Selection and estimation in PSU 6**

The value of $f_2$ is

$$f_2 = \frac{f}{f_1} = \frac{40}{3000} \times \frac{2500}{2 \times 300} = \frac{1}{18}$$

Number of households to be selected ($b_i$) will be

$$b_i = \frac{1}{18} \times 250 = 14$$

Since this is lower than 16 (the lower limit), 16 households are to be selected in PSU 6. The overall probability of selection of a household therefore would be

$$= \frac{2 \times 300}{2500} \times \frac{16}{250} = 0.015360$$

The point to remember is that a straightforward desirable design is one that is self-weighting and has clusters of equal size (number of households selected in a PSU). The two generally does not go hand-in-hand and one needs to balance the departures from i) self-weighting nature of a design and ii) having equal cluster size.

Being self-weighting means the design is unbiased. The variation in cluster size increases the sampling variance. Controlling the variations facilitates the estimation process. As mentioned, it is good to have the coefficient of variation of cluster size (equation 3.19) less than 0.1.

#### 10.3.3.4 An alternative two-stage selection

As a variant of the above method, similar two-stage selection can be employed in which equal number of households are selected from each selected village. For example, one could consider selecting 30 households in each selected village. The probability of selecting $j$th household in $i$th village will be

$$= \frac{a \times V_i}{V} \times \frac{30}{V_i'}$$

Here, the advantage is having a fixed sample take from each village. Unlike the earlier case, the design here will not be self-weighting. The probability of selection will vary. Theoretically, however, the probability of selection of a household is not expected to vary much. The ratio $\frac{V_i}{V_i'}$ will depend upon the differential rate of growth of the villages, which is not likely to differ much. The fluctuations can be heavy if there are errors in enumeration in the village sizes. If there is an error in $V_i$, and it is required to keep the sample size fixed at 30 households, it will not be possible to make any adjustment to control the variations in the departure from an EPSEM. In the earlier case, considered in NFHS, the basic design remains self-weighting and keeping track of the variation in the sample selected in a PSU was relatively easy.

### 10.3.4 Design for Urban Area

The selection of households in urban areas had been done in three stages. Wards, the lowest administrative area in urban India, were considered as the PSU. All the urban wards of a state were arranged according to their geographical region and by female literacy. Required number of wards was then selected using the PPS sampling. The second stage of selection consisted of selection of one census enumeration block (CEB) from a selected ward. Although the size of the CEBs was likely to be more or less same, wherever possible, a PPS procedure was adopted to select a CEB. Listing of households in the selected CEB provided the necessary frame for the selection of households. The selection of households in a selected CEB was done in a manner similar to the rural areas. The selection probabilities are defined as follows:

$$f = \text{overall probability or the sampling fraction}$$
$$= \frac{\text{Sample size or no.of units to be selected in urban}}{\text{Urban population or no.of urban units at survey time}}$$

$$f_1 = \text{probability of selecting a ward}$$
$$= \frac{a' \times \text{population of the selected ward}}{\text{Urban population (total) considered for selection of wards}}$$

where, $a'$ is the number of wards selected

$$f_2 = \text{probability of selecting a CEB in a selected ward}$$
$$= \frac{\text{population of selected CEB}}{\text{Population of selected ward where CEBs is selected}}$$

$$f_3 = \text{probability of selecting an ultimate unit (household or women)}$$

The present design considered that

$$f = f_1 \times f_2 \times f_3$$

Since the values of $f, f_1$ and $f_2$ were estimated or available, $f_3$ was obtained as

$$f_3 = \frac{f}{f_1 \times f_2}$$

Although a ward was a large unit for a PSU, there were few points worth considering in its favour. Wards have a well-defined boundary while CEBs lack this property. It is often difficult to locate the boundary of a CEB, particularly, if sufficient time has elapsed since the last census. One disadvantage of considering ward as PSU is that it becomes a three-stage design. However, since one CEB is generally selected from each selected ward, it can be, for all practical purposes, considered as a two-stage design.

### 10.3.5 Estimation

This section discusses about the weights required for estimating different parameters.

#### 10.3.5.1 Computation of weights

A first step for estimation of any parameter from a survey is computation of weights (Kalton and Flores-Cervantes 2003, Kish, 1992), if it is not a self-weighting design. As mentioned before, inverse of probability of selection of a unit forms its weight, also known as the base weight. The base weight can be modified to take into account the coverage and/or response errors. Computation of weights has been discussed in Chapters 2 and 3 and the adjustment for non-response is illustrated with an example (Example 3.5). The computation of weights in NFHS was quite straightforward, particularly when the basic structure of a design in a state was self-weighting. For example, within a domain, that is, within rural or urban areas in a state, in NFHS, selection of units was generally done in such a way that the design remained self-weighting. This was only violated if estimates were required for some cities in a state or in other words, if some cities in a state were also regarded as independent 'domains'. The departure from being a self-weighting design also occurred when proportional allocation was not employed in stratification. As mentioned earlier, this was done, whenever required to facilitate obtaining more cases of HIV/AIDS in NFHS-3.

However, the basic requirement for estimation of weights is to have the probability of selection of a unit. If two units have same probability of selection as well as the extent of non-response, they will have same weights. However, in NFHS, since another stage of selection was involved for obtaining information on domestic violence experienced by a woman, in order to estimate prevalence of domestic violence, a separate set of weights were needed to be computed. Similar was the case for estimating proportion of women or children who were anaemic. Since the testing of blood induced another layer of non-response in the data, a separate set of weights needed to be computed for such estimation.

In NFHS, all eligible women in a selected household had been included in the sample. Therefore, the probability of selection of women was same as that of the household in which she resided. However, since response rates were different (Chapter 5) for different groups of women they would have different weights.

All the clinical anthropometric and biochemical (CAB) components, including HIV testing for which there was a separate consent form, the response rate was likely to be different and would have separate weights. The data file, in addition to household weights, also has weights for eligible women to estimate parameters that are based on them, one for women selected for questions on domestic violence, and one for estimating all the components of CAB.

#### 10.3.5.2 Estimation of parameters

The estimation of a mean ($\bar{y}$) of a parameter is obtained as usual:

$$\bar{y} = \frac{\sum_i w_i y_i}{\sum_i w_i}$$

The estimation of sampling variance of an estimate is usually done by using the Taylor linearization approach (equations 3.15–3.18). In NFHS, in order to select PSUs, the design used PPS sampling without replacement, and hence, an unbiased estimate of the sampling variance of parameter of interest was not possible. A good approximation of it can be obtained by using the Taylor linearization approach. Statistical packages, such as STATA or SPSS can be used to estimate both mean and variance of a parameter as demonstrated below.

Example 10.2

Here we use National Family Health Survey-2 (1998-99) data from the Indian state of Tamil Nadu to illustrate the calculation of weighted estimates and standard error of the estimate. We also show STATA command on how to calculate 'design effect' often reported in a large-scale survey.

The data used in this example can be downloaded free from the website: www.measuredhs.com after a free user registration and completion of a data user form.

We compute the contraceptive prevalence rate for currently married women aged 15-49 in rural Tamil Nadu for the year 1998-99. A variable (v312) on 'Method currently using' is available in the data set and is recoded as 0 for 'not using' and 1 for 'using'. The data are restricted to currently married rural women (use v501 'marital status' and v025 'place of residence' to select data for rural currently married women).

First, we need to specify the sample design and sample weight. In order to specify a two-stage stratified design and probability weight the following codes may be used in STATA:

```
svyset, clear /*clearing any previous survey setup*/
svyset v001 [pw = wt], strata(s022) || _n .
```

where v001 is the variable denoting PSU, wt is the probability weight of each unit [wt = sv005/1000000, where sv005 is the state individual women weight included in the data set. Weight variables are generally saved as a 7-8 digit integer for precision as well as efficient use of digital space. One needs to divide them by 1000000 before using], s022 is the variable available in the data set describing stratification used at the first stage selection.

The code above specifies a two-stage design: at the first stage, PSUs (v001) are selected using a stratification defined by the variable s022. The second stage is defined by ||_n and is interpreted as random sampling at the second stage.

The output of the above code look like below:

```
. svyset v001 [pw = wt], strata(s022) || _n
      pweight: wt
          VCE: linearized
  Single unit: missing
     Strata 1: s022
         SU 1: v001
        FPC 1: <zero>
```

Note: It's a good practice to provide fpc for the first stage even if its value is small. In case you don't specify a fpc STATA will assume that the first stage selection was with replacement and consequently the second stage ( ||_n) will be ignored. To specify fpc, following codes should be used:

```
. svyset v001 [pw = wt], fpc(fpc1) strata(s022) || _n
```

where fpc1 is the variable denoting finite population correction.

*Calculating CPR [CurUse is a recoded variable]

STATA code and output

```
. svy: mean CurUse
(running mean on estimation sample)

Survey: Mean estimation

Number of strata =      23          Number of obs     =     2305
Number of PSUs   =      88          Population size   = 2748.09
                                    Design df         =       65

--------------------------------------------------------------
             |             Linearized
             |       Mean   Std. Err.     [95% Conf. Interval]
-------------+------------------------------------------------
      CurUse |   .4883754   .0117013      .4650064    .5117444
--------------------------------------------------------------
```

To calculate design effect:

```
. estat eff, deft

----------------------------------------------
             |             Linearized
             |       Mean   Std. Err.      DEFT
-------------+--------------------------------
      CurUse |   .4883754   .0117013    1.12362
----------------------------------------------
```

## 10.4 OTHER GLOBAL LARGE-SCALE SURVEYS

Other than DHS, which collects information on health, HIV and nutrition and have been undertaken in more than 70 countries, there are quite a few other large-scale surveys that are being carried out globally with different objectives. The Living Standard Measurement Survey (LSMS), Life in Transition Survey (LITS), and Global Adult Tobacco Survey (GATS) are some of them. The LSMS provides household-level data on living condition in largely developing countries and includes variety of aspects, such as consumption behaviour, employment pattern and environmental condition, depending on interest of the country where it is being conducted. LITS, on the other hand, is carried out in 16 countries in central and eastern Europe and the Baltic states, south-eastern Europe, and the Commonwealth of independent states and Mongolia. Its objective is to understand how life in transition in an open-market-related economy has been shaping in terms of poverty and inequality, satisfaction of people with public services, etc. GATS is a global survey, conducted in various countries, with a focus on understanding tobacco use and exposure of people to smoking.

All the above surveys are household-based surveys where the unit of study is a household. Majority of them require selection of individuals, generally adult females or males in a selected household. The main structure of the sample design used in all these surveys is similar, though the specific design adopted varies according to the type of survey and the country where it is to be implemented. All of them employ a multistage probability sampling design to select households or individuals within a selected household. Following decisions are noteworthy that could distinguish them from each other.

- **Constitution of PSUs**: They are appropriate geographically defined area units, in a country. Generally, at the time of census, small areas consisting of around 200 households are delineated. These areas have distinct and identifiable boundaries and are often called enumeration areas (EAs). These EAs are fruitfully utilized as the PSUs. Some countries like Kenya, Nigeria, the Philippines, etc. have 'master samples of EAs' to facilitate conducting sample surveys (discussed subsequently).
- **Mode of stratification of PSUs**: The geographical location of PSUs, in the absence of any other relevant variables, is invariably used to stratify them. The size of PSUs is also recommended and used, if practicable.
- **Selection of PSUs**: When PSU sizes are available, selection of PSUs is generally done by PPS. The recent population census is an important source of obtaining estimated size of PSUs.
- **Listing**: Accurate and up-to-date listing of dwellings/households is done in each selected PSU to facilitate selection of households.
- **Selection of households**: Systematic sampling is generally used for this. Selection of individuals from a selected household, if necessary, is often done assigning equal probability to each of them.

### 10.4.1 Use of Master Sample in Survey Designs

A variation in the sampling design adopted can arise due to the use of a master sample (MS). To facilitate sample surveys, many countries select and maintain a large list of areal units, known as MS. The areal units are usually such that they can be used as a PSU for a future survey and are often EAs. At the time of a census, a large number of EAs are selected and kept for record along with their probability of selection. Each EA has distinct and non-overlapping boundaries that can be identified. The number of EAs selected are usually large, enough to support the selection of quite a few sample surveys during an intercensal period. Future surveys can use the MS for the PSU selection and households are then selected from each selected PSU, either in two or more stages.

Regular listing may be carried out in all the PSUs in a MS to obtain their updated household size. This can help in obtaining regular information on parameters of socio-economic changes occurring in a population. In addition, it can reduce the cost of listing in future surveys, which can be a distinct advantage in using an MS.

To obtain the probability of selection of a household from a population into the sample, and in our endeavour to have an EPSEM design, the sample selection needs to be done in a specific manner. Let us first define the following probabilities:

$P_i$ = probability of selecting *i*th EA in a population into the MS

$P_i'$ = probability of selecting the *i*th EA, which is selected in the MS, into the present sample

$P_{ij}'$ = probability of selecting *j*th household from *i*th EA selected from MS in the sample

Since the interest is on estimating the population parameter, the probability of selection of (*i*,*j*)th household of the population ( $P_{ij}''$ ).

$$P_{ij}'' = P_i \times P_i' \times P_{ij}'$$

Further suppose,

$H_i$ = number of households in *i*th EA in the population,
$E$ = number of EA in the population,
$e$ = number of EA selected in MS,
$a$ = number of EA selected in the sample.

Suppose, PPS is used to select EAs both the times, that is, to select EA from the population for the MS as well as to select in the sample from the MS, and an equal number of households (*k*) are selected by an EPSEM at the last stage, then

$P_i = \dfrac{e \times H_i}{H}$, where *H* is the total household size in the population,

$P_i' = \dfrac{a \times H_i}{H'}$, where *H′* is the total household size of all EAs in the MS, and

$P_{ij}' = \dfrac{k.}{H_i}$

It can be seen that $P_{ij}''$, in this case, will not be constant and the design will not be self-weighting. If *a* PSUs, in the sample, are selected by an EPSEM, (that is, if $P_i' = \frac{a}{e}$ ) it will be self-weighting (assuming that there is no increase or decrease in the size of EAs at the time of selecting them for MS and the survey). But if EAs are selected in MS with an EPSEM, that is, $P_i = \frac{e}{E}$ then the selection of EA into the sample $(P_i')$ should be by PPS to make it self- weighting.

### 10.4.2 Example of GATS sample design in Nigeria

In Nigeria, the GATS was conducted in 2012, using its MS. Its sampling design is briefly discussed (for detailed discussion, please refer to https://www.google.com/url?sa=t&rct=j&q=&esrc=s&source=web&cd=3&cad=rja&uact=8&ved=0CCwQFjAC&url=http%3A%2F%2Fwww.nigerianstat.gov.ng%2Fpages%2Fdownload%2F157&ei=kAVOVfHlLJCwuASAoIDADw&usg=AFQjCNF7lYdSUu2i7GRKZl0TJkoMs32jyw&sig2=kQAFFej2Wch-ZqgPetlZdA&bvm=bv.92885102,d.c2E, accessed on 9 May 2015). The target population was all men and women of age 15 years or more who are living in households. The census in Nigeria was conducted in 2006. It formed the basis for constructing its MS through the National Integrated Survey of Households. The MS consisted of 7400 EAs which were selected systematically from the list of EAs arranged according to the states. From every state, 200 EAs were selected to form the MS. EAs have an average of 37 households in urban and 33 in the rural areas. A total of 1100 EAs were selected for the GATS sample. They were distributed to each and every state in proportion to their population. The selection of EAs was done with PPS that yielded 300 EAs in urban and 800 in rural areas. Since estimates were needed for both urban and rural areas, approximately equal number of households were needed in both the areas. Assuming a differential level of non-response in the two areas, 18 and 7 households were selected in urban and rural areas, respectively. The selection of households was done systematically, after conducting a detail household listing in the selected EAs. Selection of individual adults (female or male) was carried out by assigning each household an equal chance to be either a female or a male household. This was done using a handheld device.

## 10.5 NATIONAL SAMPLE SURVEYS (NSS) IN INDIA

The Government of India, set up the National Sample Survey Organization (NSSO) in 1950. Its primary aim has been to conduct surveys on a regular basis to obtain information on a variety of socio-economic, health and demographic characteristics of population of the country. The sampling design adopted in these surveys differs slightly from the other large-scale surveys in the field. As an example, let us discuss some specific information about the 60th round of the survey that was on household consumer expenditure, employment and unemployment and morbidity and healthcare (NSSO, 2004). The essence of the sample design used in the survey is outlined briefly here. The basic feature of the survey is to have two different subsamples drawn independently. The estimates thus derived from the two subsamples are particularly helpful in estimation of the sampling variance.

This household-based survey was conducted during January–June 2004 to cover the entire country. The total sample size was equally subdivided into two subsamples that were covered during January–March and April–June of the year.

In rural areas, villages formed the PSUs (it is referred to as the first stage unit, FSU) which were selected by PPS with replacement.[1] Before their selection, villages were suitably stratified considering their size and geographic location. The second stage of selection was mainly of households. For smaller size villages (less than around 1200 population or about 250 households), households were selected by the SRS. The larger villages were first subdivided into three or more hamlets (segments). Two hamlets were selected by SRS, and then the households' selection was done in a similar manner.

In urban areas, the NSSO has blocks, similar to the CEB, known as Urban Frame Survey (UFS) blocks. The selection of households was similar to that in the rural areas. That is, while UFS were selected by PPS with replacement, the households either directly or after the selection of sub-blocks were carried out by SRS. The two subsamples selected independently provides two separate estimates of a population parameter. The estimated mean or proportion or total of a parameter, within each stratum, will be a simple average of the two independent estimates from the subsamples. For the overall estimate, these stratum level means can be combined by the usual procedure of stratified sampling (equation 3.32). The variance can be computed by using their difference within each stratum and following the similar averaging procedure (equation 3.33).

For detailed report please visit: http://mospi.nic.in/Mospi_New/site/inner.aspx?status=4&menu_id=53, accessed on 9 May 2015.

# PART II: EVALUATION DESIGN

## 10.6 ILLUSTRATION OF EVALUATION DESIGNS

Few illustrations of evaluation designs implemented in practice have been summarized below.

### 10.6.1 Impact Evaluation of a Life Skills Education Intervention on Adolescent Girls' Empowerment

A research study was undertaken to assess the extent to which Centre for Development and Population Activities, USA (CEDPA) life skills education programme: the Better Life Options Program (BLP) for Adolescent Girls in India can empower adolescent girls and address the vulnerabilities they face. Specifically, the study aimed to assess the extent to which participation in the above intervention enhanced girls' ability to overcome the vulnerabilities they face and make successful transition to adulthood.

The intervention was conducted in the rural areas of two blocks of Lucknow district of Uttar Pradesh state of India: Malihabad and Maal. Eighteen villages of Malihabad served

[1] In Chapter 2 (Section 2.5.1), selection of clusters systematically by PPS using a single random number has been elaborated. The procedure refers to selection by PPS without replacement, that is, a same cluster cannot be selected more than once. For the selection by PPS with replacement, separate random numbers, instead of the interval, can be drawn and used for each of the selection.

as intervention sites and the nine villages of Maal as control sites. The intervention was implemented by the NGO Prerana during the period August 2006 to October 2007.

### Study design

In total, 18 census villages in Malihabad were identified for intervention activities. The intervention block was identified by the implementing agency and was preferred because of the agency's prior presence in the block. The selection of villages was done to ensure the following aspects: 1) ensure geographical contiguity of the intervention villages (a condition desirable for the implementing agency) and 2) ensure a sufficient number of adolescent girls aged 13–17 years – the focus of the project – available for intervention and evaluation. In a similar way, nine contiguous villages in Maal block were selected as control area and this area was selected in such a manner that it matched with the intervention area as closely as possible and this matching was done using a limited data available from the census. The average distance between the intervention area and the control area was approximately 30 km, ensuring that no diffusion of programme occurred from the intervention to the control area.

### Evaluation design

In order to rigorously evaluate the effect of the intervention on the girls' life, a quasi-experimental research design (evaluation design with matching; discussed in Section 7.4.3.1) was adopted, with panel surveys undertaken in the control and intervention villages prior to the implementation of the Choose a Future programme (baseline) and at its conclusion (end line).

As a first step, a complete listing of all unmarried girls aged 13–17 years in 27 villages was done and the baseline survey was administered to all of them: both in the intervention and control sites. All eligible girls were invited to participate in the intervention that lasted for 15 months.

Since a panel design was adopted, in order to retain the sample for the end line, special efforts were made to track all baseline respondents in the inter-survey period. Three visits were made to update respondents' addresses and contact details and records of their marriage and migration status were kept. We would like to emphasize here that while tracking of baseline respondents is desirable, it is not necessary, particularly since such efforts are resource intensive.

The end-line panel survey was conducted 2 years following the baseline survey, and some 9–15 months following the conclusion of the intervention. Investigators used addresses collected at the time of baseline as well as tracking records in order to visit households and contact baseline survey respondents. The end-line survey was conducted between late August and early October 2008 and covered girls now aged 15–19 years.

The baseline response rates were over 80 per cent and follow-up rate was 94 per cent, giving much desired validity to the panel data collected. At the baseline, both intervention and control areas had similar socio-economic characteristics, giving validity to our design and subsequent analysis of impact of the programme.

### Possible alternative design

It was possible to adopt an even more rigorous alternative design, such as cluster-randomized control design. One could take up one block and randomly assign required number of villages to treatment or control arms. While one can ensure a close matching between the arms by adopting a restricted randomization scheme, it would be difficult to control contamination between control and intervention villages if they happen to be in close vicinity. Controlling contamination would have required an expensive mapping exercise and an increase in cost of intervention due to geographical spread of the intervention area. Given the resources available, it was decided to adopt a quasi-experimental design, which is the next best design possible in the situation.

### Impact analysis

The study used DiD method to calculate impact. The DiD method takes care of secular changes in both the treatment populations and assumes that any changes external to the intervention was similar for both. As discussed before the samples were similar in their socio-demographic distributions and in their baseline levels of key indicators in question. Findings indicate that participation in the intervention did have a positive net effect on the adolescent girls' agency (involvement in decision making, mobility, sense of self-efficacy and access to economic resources), gender role attitudes and awareness of various sexual and reproductive health matters. However, it was observed that statistically significant impact was visible only among those adolescents who had participated in the intervention regularly (Acharya et al., 2009).

## 10.6.2 Fisher's Tea Test

Fisher's tea test is a classic experiment in which units were assigned randomly to the two groups $T$ and $C$ (Fisher, 1935). The experiment was to test the claim of a lady that she could specify, by testing tea, whether milk or the liquor was added first to prepare it. The unit of observation was a cup of tea, and the two types of preparation were cups of tea in which liquor was poured first and then the milk, say, the treatment ($T$) and tea in which milk was poured first and subsequently the liquor ($C$). Few disturbing variables, in the experiment, could be type of tea leaves used, particular brand of milk used, temperature of the liquor, whether milk used was hot or cold, etc. The experiment was, however, conducted disregarding disturbing variables. The observations on $T$ and $C$ were presented to the lady tester in a random manner. That is, instead of matching, only the sequence of presentation of a cup of tea was randomized. A total of eight cups, four of the each type, were thus presented to the tester. The output observed was whether the tester's specification agreed with the actual preparation. It was, therefore, a dichotomous variable denoting whether the tester was right (say, $y = 1$) or wrong ($y = 0$). Under the null hypothesis that the answer was given randomly (that is, the tester had no specific skills of detection), the tester had a 50 per cent chance of giving a right answer. Applying the exact binomial test (Section 9.5.2.1), it is possible to test the hypothesis.

### 10.6.3 Impact of an Intervention to Promote Use of Intra Uterine Device in a Population

To promote the use of IUD as a spacing method of family planning, an intervention study was conducted in the state of Gujarat, India. The intervention was mainly aimed at educating and improving the technical skills of service providers so that delivery of spacing method services can be improved. In addition, provisions of IUD services were provided in selected public health facilities.

The study adopted a before-after design in rural and urban areas of Vadodara district of the state. Although quite a few variables were considered for understanding the impact (Khan et al., 2008), three specific outputs are illustrated here. They are:

**(i)** proportion of providers having poor knowledge about IUD,
**(ii)** proportion of IUD users who reported quality of IUD services to be good and
**(iii)** average number of IUD insertion.

For the impact estimation, observations were made on the service providers. All the providers in the experimental area were interviewed twice, at the baseline (before introducing the intervention) and after 9 months (end line). In addition, 10 IUD users from each study area were randomly selected (based on registers obtained from health care providers) at the said two points of time. The sample consisted of 83 IUD users at the baseline and 203 at the end line. The reason for the differences in the number of IUD users was mainly because of large percentage of false cases at the baseline (though the register showed them as IUD users, they were found to have never used when contacted). Information on the quality of IUD services was collected from the users. Data on number of IUD insertions were obtained from the service statistics.

The impact was as usual computed by taking the difference in the estimated value of the three parameters at baseline and end line.

### 10.6.4 An Experiment to Test Effectiveness of a Medicine

An experiment was conducted to test toxicity and effectiveness of administering one daily dose of tobramycin (*T*) as against three doses daily of the same antibiotic (*C*) to treat febrile neutropenic children undergoing stem cell transplantation. It was a treatment-control design in which children admitted to hospitals were first tested for the eligibility of their participation in the trial. The test involved examination of their physical condition as well as the required consent to participate (Sung et al., 2003).

In total, 60 children were screened as eligible and were randomly assigned to *T* and *C*. There were 29 cases in *T* and 31 in *C*. The eligible children were stratified into six groups according to their age (three categories) and stem cell donor (two categories).

It was carried out as a double-blind experiment. Each dose, prepared by the pharmacy staff, was given in a 50-ml syringe. Every eligible child received three injections. For those in the *T* group, only one injection contained the actual dose of tobramycin and the other two were normal saline. The members of the research team were unaware of the specific schedule of the administration of the doses.

In the initial stages of the experiment, before the allocation of eligible children, during the clinical evaluation to decide their eligibility, measurements were taken on outputs (nephrotoxicates) of each child. Since the units were assigned randomly to *T* and *C*, it was not necessary to use the measurements done at the baseline. A comparison between the two outputs, using an appropriate statistical test, would suffice to understand the impact of *T*, if any. However, since the sample size was not quite large, an adequate representation of children according to age and the donor type might be lacking. Hence, the baseline information was used, as a safeguard, to obtain the impact by the DiD approach (mentioned in Section 7.4.3.1).

### 10.6.5 An Intervention to Reduce Post-Partum Haemorrhage

An experiment was conducted, in a multi-country project, to examine the effect of few interventions to reduce post-partum haemorrhage, which happens to be an important cause of maternal morbidity and mortality. The intervention was a particular medicine, Misoprostol that can be administered orally and is not expensive (Gülmezoglu et al., 2001).

The objective was to test its effectiveness in terms of reducing the blood loss of 1000 ml or more. In total, 18,530 women were found eligible for the trial. Among those who were admitted to labour wards, the eligibility was based on their health condition, if the delivery was not planned for caesarean section and if agreed to participate in the trial. Women were randomly assigned to the two groups: i) who received misoprostol, *T* (9264 women) and ii) who were given the usual oxytocin, *C* (9266 women).

This is an evaluation design with randomization of units to treatment and control arm (7.4.2). The difference in the outputs (extent of blood loss) between the two groups, *T* and *C*, provided the necessary effect.

# References

Acharya, R. S. Kalyanwala, and S. J. Jejeebhoy. 2009. *Broadening Girls' Horizons: Effects of a Life Skills Education Programme in Rural Uttar Pradesh*. New Delhi: Population Council.

Andridge, Rebecca R., and Roderick J. A. Little. 2010. 'A Review of Hot Deck Imputation for Survey Non-response.' *International Statistical Review* 78(1): 40–64.

Angrist, Joshua D., Guido W. Imbens, and Donald B. Rubin. 1996. 'Identification of Causal Effects Using Instrumental Variables.' *Journal of the American Statistical Association* 91(434): 444–55.

Bethlehem, Jelke. 2009. 'The Rise of Survey Sampling.' *Discussion Paper (09015)*. The Hague/Heerlen: Statistics Netherlands.

Boyd, Harper W. Jr., Ralph Westfall, and Stanley F. Stasch. 1989. *Marketing Research: Text and Cases*. Boston, MA: Irwin.

Breslow, N. E., and N.E. Day. 1980. *Statistical Methods in Cancer Research*. Vol. I. International Agency for Research on Cancer (IARC). Lyon, France.

Brick, J. M., and G. Kalton. 1996. 'Handling Missing Data in Survey Research.' *Statistical Methods in Medical Research* 5: 215–38.

Campbell, D. T. 1957. 'Factors Relevant to the Validity of Experiments in Social Settings.' *Psychological Bulletin* 54(4): 297–312.

Campbell, D., and J. Stanley. 1963. *Experimental and Quasi-Experimental Designs for Research*. Chicago: Rand McNally.

Cochran, W. G. 1950. 'The Comparison of Percentages in Matched Samples.' *Biometrika* 37: 256–66.

———. 1961. 'Comparison of Methods for Determining Stratum Boundaries'. *Bulletin of the International Statistical Institute* 37(2): 345–58.

———. 1977. *Sampling Techniques*. 3rd edition. New York: John Wiley & Sons.

Dalenius, T., and J. L. Hodges. 1959. 'Minimum Variance Stratification.' *Journal of the American Statistical Association* 54: 88–101.

Do, Y. K., and K. Y. Wong. 2012. 'Awareness and Acceptability of Human Papillomavirus Vaccine: An Application of the Instrumental Variables Bivariate Probit Model.' *BMC Public Health* 12: 31. doi:10.1186/1471-2458-12-31

Edwards, A. L. 1948. 'Note on the "Correction for Continuity" in Testing the Significance of the Difference Between Correlated Proportions.' *Psychometrika* 13: 186–7.

Ekman, G. 1959. 'An Approximation Useful in Univariate Stratification.' *Annals of Mathematical Statistics* 30: 219–29.

Fisher, R. A. 1925. *Statistical Methods for Research Workers*. Edinburgh, UK: Oliver and Boyd.

——— 1935. *The Design of Experiments*. Edinburgh, UK: Oliver and Boyd.

Fisher, R. A., and F. Yates. 1963. *Statistical Tables for Biological, Agricultural and Medical Research.* England: Longman Group Limited.

Fleiss, Joseph L. 1973. *Statistical Methods for Rates and Proportions.* p. 79. New York: John Wiley & Sons.

Freedman, David. (1999). 'From Association to Causation: Some Remarks on the History of Statistics.' *Statistical Science* 14(3): 243–58.

Goodman, Roe, and Leslie Kish. 1950. 'Controlled Selection- A Technique in Probability Sampling.' *Journal of the American Statistical Association* 45: 350–72.

Graunt, J. 1662. *Natural and Political Observations upon the Bills of Mortality.* London: Martyn.

Gülmezoglu, A. Metin, José Villar, Nguyen Thi Nhu Ngoc, Gilda Piaggio, Guillermo Carolli, Lekan Adetoro, Hany Abdel-Aleem, Linan Cheng, G. Justus Hofmeyr, Pisake Lumbiganon, Christian Unger, Walter Prendiville, Alain Pinol, Diana Elbourne, Hazem El-Refaey, and Kenneth F. Schultz. 2001. 'WHO Multicentre Randomised Trial of Misoprostol in the Management of the Third Stage of Labour.' *Lancet* 358: 689–95.

Guo, Shenyang, and Mark W. Fraser. 2010. *Propensity Score Analysis.* California: Sage Publications.

Hagood, Margaret Jarman, and Elaneanor H. Bernert. 1945. 'Component Indexes as a Basis for Stratification in Sampling.' *Journal of the American Statistical Association* 40(231): 330–41.

Hansen, M. H., T. Dalenius, and B. J. Tepping. 1985. 'The Development of Sample Surveys of Finite Populations.' In *A Celebration of Statistics*: The ISI Centenary Volume, 327–354. Berlin: Springer-Verlag.

Hansen, Morris H., William N. Hurwitz, and William G. Madow. 1953. *Sample Survey Methods and Theory.*Vol.1. New York: John Wiley & Sons.

Hayes, Richard J., and Lawrence H. Moulton. 2009. *Cluster Randomised Trials.* New York: CRC Press, Taylor and Francis Group.

Holland, Paul W. 1986. 'Statistics and Causal Inference.' *Journal of the American Statistical Association* 81(396): 945–60.

Kalsbeek, William D. 1986. 'Nomad Sampling: An Analytic Study of Alternative Design Strategies.' *Proceedings of the Section on Survey Research Methods, American Statistical Association*: 164–69. (Accessed April 29, 2015. http//www.amstat.org/sections/srms/Proceedings/papers/1986_028.pdf)

Kalton, Graham. 2001. 'Practical Methods for Sampling Rare and Mobile Populations'. *Proceedings of the Annual Meeting of the American Statistical Association*, Chicago, Illinois, August 5–9.

———. 2009. 'Methods for Oversampling Rare Subpopulations in Social Surveys.' *Survey Methodology* 35(2): 125–41.

Kalton, Graham., J. Michael Brick., and Thanh Lê. 2005. 'Estimating Components of Design Effects for Use in Sample Design.' In *Household Surveys in Developing and Transition Countries: Studies in methods.* F(96), 95–122. New York: Department of Economic and Social Affair, United Nations.

Kalton, Graham, and Ismael Flores-Cervantes. 2003. 'Weighting Methods.' *Journal of Official Statistics* 19(2): 81–97.

Kalton, Graham, and Howard Schuman. 1982. 'The Effect of Question on Survey Responses: A Review.' *Journal of the Royal Statistical Society. Series A (General)* 145(1): 42–73.

Kanouse David E., Sandra H. Berry, and Naihua Duan. 1999. 'Drawing a Probability Sample of Female Street Prostitutes in Los Angeles County.' *Journal of Sex Research* 36(1): 45–51.

Kempthorne, O. 1952. *The Design and Analysis of Experiments.* New York: John Wiley & Sons.

Khan, M. E., S. S. Kar, V. K. Desai, P. Patel and B. P. Itare. 2008. *Increasing the Accessibility, Acceptability and Use of the IUD in Gujarat, India.* New Delhi: Population Council.

Kish, Leslie. 1949. 'A Procedure for Objective Respondent Selection within the Household.' *Journal of the American Statistical Association* 44(247): 380–7.

———. 1987. *Statistical Design for Research.* New York: John Wiley & Sons.

———. 1991. 'Taxonomy of Elusive Populations.' *Journal of Official Statistics* 7(3): 339–47.

———. 1992. 'Weighting for Unequal $P_i$.' *Journal of Official Statistics* 8(2): 183–200.

———. 1995. *Survey Sampling*. New York: John Wiley & Sons.

———. 2003a. 'The Hundred Years' of Wars of Survey Sampling.' In *Leslie Kish: Selected Papers*, edited by Kalton Graham and Steven Heeringa, 5–20. New Jersey: John Wiley & Sons.

———. 2003b. 'Some Statistical Problems in Research Design.' In *Leslie Kish: Selected Papers*, edited by Kalton Graham and Steven Heeringa, 21–34. New Jersey: John Wiley & Sons.

Kish, Leslie, and Dallas W. Anderson. 1978. 'Multivariate and Multipurpose Stratification.' *Journal of the American Statistical Association* 73(361): 24–34.

Kish, L., and M. R. Frankel. 1974. 'Inference from Complex Samples.' *Journal of the Royal Statistical Society (B)* 36: 1–37.

Kish, Leslie, W. Lovejoy, and P. Rackow. 1961. 'A Multi-Stage Probability Sample for Continuous Traffic Surveys.' *Proceedings of the Social Statistics Section, American Statistical Association*. 227–30. Washington, D.C: American Statistical Association.

Kitagawa, Evelyn M. 1964. 'Standardized Comparisons in Population Research'. *Demography* 1(1): 296–315.

Lahiri, D. B. 1958. 'Observations on the Use of Interpenetrating Samples in India.' *Bulletin of the International Statistical Institute* 36(3): 144–52.

Lemeshow, Stanley, David W. Hosmer Jr., Janelle Klar, and Stephen K. Lwanga. 1990. *Adequacy of Sample Size in Health Studies*. New York: John Wiley & Sons.

Maddala, G. S. 1977. *Econometrics*. London: McGraw-Hill.

Mahalanobis, P. C. 1946. 'Recent Experiments in Statistical Sampling in the Indian Statistical Institute.' *Journal of Royal Statistical Society* 109: 325–78.

McClave, James T., P. George Benson, and Terry Sincich. 2005. *Statistics: for Business and Economics*. Upper Saddle River, New Jersey: Pearson Prentice Hall.

McGinnis, R. 1958. 'Randomization and Inference in Sociological Research.' *American Sociological Review* 23: 408–14.

McKenzie, David J., and Johan Mistiaen. 2009. 'Surveying Migrant Households: A Comparison of Census-Based, Snowball and Intercept Point Surveys.' *Journal of Royal Statistical Society, A* 172(2): 339–60.

McKinlay, Sonja M. 1977. 'Pair-Matching – A Reappraisal of a Popular Technique.' *Biometrics* 33: 725–35.

McNemar, Q. 1947. 'Note on the Sampling Error of the Difference between Correlated Proportions or Percentages,' *Psychometrika* 12: 153–7.

Milligan, Paul, Alpha Njie, and Steve Bennett. 2004. 'Comparison of Two Cluster Sampling Methods for Health Surveys in Developing Countries.' *International Journal of Epidemiology* 33: 469–76.

Namboodiri, N. Krishnan. 1970. 'A Statistical Exposition of the "Before-After" and "After-Only" Designs and Their Combinations.' *American Journal of Sociology* 76(1): 83–102.

National Sample Survey Organization (NSSO). 2004. *Morbidity, Health Care and the Condition of the Aged*. Report No.507 (60/25.0/1), New Delhi: NSSO.

O' Muircheartaigh, C. A., and S. T. Wong. 1981. 'The Impact of Sampling Theory on Survey Sampling Practice: A Review.' *Bulletin of the International Statistical Institute* 49(1): 465–93.

Pagano, Marcello, and Gauvreau Kimberlee. 2000. *Principles of Biostatistics*. Pacific Grove, California: Duxbury.

Potthoff, Richard F., Susan Halabi, Joelen M. Schildkraut, and Beth Newman. 2008. 'Flexible Frames and Controlled Sampling in Case-Control Studies: Weighters (Survey Statisticians) Versus Anti-Weighters (Epidemiologists)'. *American Statistical Association* 62(4): 307–13.

Ram, F., and T. K. Roy. 2004. 'Comparability Issues in Large Sample Surveys: Some Observations.' In *Population, Health and Development in India: Changing Perspective*, edited by T. K. Roy, M. Guruswamy, and P. Arokiaswamy, 40–56. New Delhi: Rawat Publication.

Rao, J. N. K., H. O. Hartley, and W. G. Cochran. 1962. 'On a Simple Procedure of Unequal Probability Sampling without Replacement.' *Journal of Royal Statistical Society, (B)* 24: 482–91.

Retherford, Robert D., and Minja Kim Choe. 1993. *Statistical Models for Causal Analysis*. New York: John Wiley & Sons.

Rosenbaum, P. R., and D.B. Rubin. 1983. 'The Central Role of the Propensity Score in Observational Studies for Causal Effects.' *Biometrika* 70(1): 41–55.

———. 1985. 'Constructing a Control Group Using Multivariate Matched Sampling Methods that Incorporate Propensity Score.' *American Statistician* 39: 33–8.

Roy, T. K., and N. Unnikrishnan Nair. 1983. 'Marriage and Marital Fertility: A Further Decomposition of Their Effects in the Study of Levels and Changes in Birth Rate'. *Canadian Studies in Population* 10: 15–30.

Roy, T. K., and Arvind Pandey. 2008. 'National Family Health Survey in India: Survey Design and Related Issues.' *Demography India* 37, Supplement, (Jan–Dec., 2008): 13–41.

Rubin, Donald B. 1974. 'Estimating Causal Effects of Treatments in Randomised and Non-Randomised Studies.' *Journal of Educational Psychology* 66: 688–701.

———. 1987. *Multiple Imputation for Nonresponse in Surveys*. New York: John Wiley & Sons.

———. 2005. 'Causal Inference Using Potential Outcomes: Design, Modeling, Decisions.' *Journal of the American Statistical Association* 100(469): 322–31.

Satterthwaite, F. W. 1946. 'An Approximate Distribution of Estimates of Variance Components.' *Biometrics Bulletin* 2: 110–4.

Singh, Abhishek, Sandip Chakraborty, and Tarun K. Roy. 2008. 'Village Size in India: How Relevant it is in the Context of Development?' *Asian Population Studies* 4(2): 111–34.

Snedecor George W., and Willam G. Cochran. 1989. *Statistical Methods*. 8th edition. Iowa: Iowa State University Press.

Solomon, R. W. 1949. 'An Extension of Control Group Design.' *Psychological Bulletin* 46: 137–50.

Som, Ranjan Kumar. 1996. *Practical Sampling Techniques*. New York: Mercel Dekker.

Stevens, W.L. 1952. 'Samples with the Same Number in Each Stratum.' *Biometrika* 39: 414–6.

Stueve Ann, Lydia N. O'Donnell, Richard Duran, Alexi San Doval, and Juliana Blome. 2001. 'Time-Space Sampling in Minority Communities: Results with Young Latino Men Who Have Sex with Men.' *American Journal of Public Health* 91(6): 922–6.

Sudman, Seymour. 1980. 'Improving the Quality of Shopping Center Sampling.' *Journal of Marketing Research* 17: 423–31.

———. 1983. 'Response Effects to Behavior and Attitude Questions.' In *Statistical Methods and the Improvement of Data Quality*, edited by Tommy Right, 85–114. Orlando: Academic Press.

Sung, Lillian, L. Lee Dupuis, Bonnie Bliss, Anna Taddio, Mohamed Abdolell, Upton Allen, Martha Rolland, Agnes Tong, Tracey Taylor, and John Doyle. 2003. 'Randomized Controlled Trial of Once-Versus Thrice-Daily Tobramycin in Febrile Undergoing Stem Cell Transplantation.' *Journal of the National Cancer Institute* 95(24): 1869–77.

Turner, Anthony G., Robert Magnani, and Muhammad Shuaib. 1996. 'A Not Quite as Quick but Much Clearer Alternative to the Expanded Programme on Immunization (EPI) Cluster Survey Design.' *International Journal of Epidemiology* 25(1): 198–203.

Unisa, Sayeed, Palepu Jagannath, Vinay Dhir, Chiranjeeva Khandelwal, Lalatendu Sarangi, and Tarun Kumar Roy. 2011. 'Population Based Study to Estimate Prevalence and Determine Risk Factors of Gallbladder Diseases in the Gangetic Basin of North India.' *The Official Journal of the International Hepato-Pancreato-Billary Association*. HPB (Oxford) 13(2): 117–125.

Verma, V. 1991. *Sampling Methods, Training Handbook.*Tokyo: Statistical Institute for Asia and the Pacific.

Verma, V., C. Scott., and C. A. O'Muircheartaigh. 1980. 'Sample Designs and Sampling Errors for World Fertility Survey.' *Journal of Royal Statistical Society, A* 143(4): 431–73.

Waksberg, Joseph. 1978. 'Sampling Methods for Random Digit Dialing.' *Journal of American Statistical Association* 73: 40–6.

Waksberg, Joseph, J. Michael Brick, Gary Shapiro, Ismael Flores-Cervantes, and Bridgett Bell. 1997. 'Dual-Frame RDD and Area Sample for Household Survey with Particular Focus on Low-Income Population.' Accessed May 9, 2015. http://www.amstat.org/sections/srms/proceedings/papers/

Warner, Stanley. 1965. 'Randomized Response: A Survey Technique for Estimating Evasive Answer Bias.' *Journal of American Statistical Association* 60(309): 63–9.

Wells, Bradley, H. 1975. 'Matching Studies', In *Measuring the Effect of Family Planning Programs on Fertility*, edited by C. Chandrasekaran and Albert I. Hermalin, 215–44. Belgium: Ordina Edition.

Wilcoxon, F., and R. A. Wilcox. 1964. *Some Rapid Approximate Statistical Procedures*. Pearl River, New York: The American Cynamid Co. Lederle Laboratories.

World Health Organization. 1991. 'Training for Mid-level Managers – Module 10: The EPI Coverage Survey,' WHO/EPI/MLM/91.10.

Worcestor, J. 1964. 'Matched Sample in Epidemiologic Studies.' *Biometrics* 20: 840–8.

Yates, F. 1934. 'Contingency Table Involving Small Numbers and the $\chi^2$ Test.' *Journal of the Royal Statistical Society* 1(2): 217–35.

Youkeles, L. H. 1963. 'Loss of Power Through Ineffective Pairing of Observations in Small Two-Treatment All-or-None Experiments.' *Biometrics* 19(1): 175–180.

Yule, G. Udny, and M. G. Kendall. 1950. *An Introduction to the Theory of Statistics*. London: Charles Griffin and Co. Ltd.

# Index